Engineering Materials

This series provides topical information on innovative, structural and functional materials and composites with applications in optical, electrical, mechanical, civil, aeronautical, medical, bio- and nano-engineering. The individual volumes are complete, comprehensive monographs covering the structure, properties, manufacturing process and applications of these materials. This multidisciplinary series is devoted to professionals, students and all those interested in the latest developments in the Materials Science field, that look for a carefully selected collection of high quality review articles on their respective field of expertise.

Indexed at Compendex (2021) and Scopus (2022)

Dzun Noraini Jimat
Editor

Deep Eutectic Solvents

Green and Sustainable Solvents

Springer

Editor
Dzun Noraini Jimat
Department of Chemical Engineering
and Sustainability
International Islamic University Malaysia
Kuala Lumpur, Malaysia

ISSN 1612-1317 ISSN 1868-1212 (electronic)
Engineering Materials
ISBN 978-981-95-1911-8 ISBN 978-981-95-1912-5 (eBook)
https://doi.org/10.1007/978-981-95-1912-5

This Springer imprint is published by the registered company Springer Nature Singapore Pte Ltd.
The registered company address is: 152 Beach Road, #21-01/04 Gateway East, Singapore 189721, Singapore

Contents

The Discovery and Development of Deep Eutectic Solvents (DESs)

Dzun Noraini Jimat, Muhammad Hasya Hasnuddin, Nurul Sakinah Engliman, Ainul Husna Abdul Aziz, and Nurul Syifa Mohd Fauzi

Abstract This chapter covers the early discovery and development of deep eutectic solvents (DESs), their physicochemical properties, and a comparison with ionic liquids (ILs). DESs and ILs share many similarities, but DESs are generally safer, cheaper, and easier to synthesize. DESs have gained popularity because they are easy to produce, non-toxic during handling, and pose minimal environmental hazards. Furthermore, DESs are better in terms of biodegradability and low toxicity, which is attributed to the natural origin of their components. As such, these solvents are considered green solvents and can serve as alternative options for industries seeking sustainable solutions. Nevertheless, both ILs and DESs offer promising prospects for further research, especially in analytical chemistry and industrial applications.

1 Introduction

Due to the growing demand for green and eco-friendly solvents, new chemical families are being explored for this purpose. Deep eutectic solvents (DESs) are among the most promising alternatives. These solvents are gaining attention because of their unique properties, which make them attractive for various industrial and environmental applications. They share many similarities with ionic liquids (ILs) but are generally safer, cheaper, and easier to synthesize. DESs are typically produced by simply mixing acids and bases, resulting in systems that contain both cationic and anionic constituents. ILs have been widely investigated and used due to their excellent physicochemical properties. ILs possess desirable properties, including low vapor pressure at ambient temperature, high thermal and chemical stability, good solubility, electrical conductivity, and recyclability.

Despite these advantages, ILs also have several drawbacks. They are costly to produce, potentially harmful to the environment, and most notably require complex synthesis and purification procedures, which limit their widespread use. Moreover,

D. N. Jimat (✉) · M. H. Hasnuddin · N. S. Engliman · A. H. Abdul Aziz · N. S. Mohd Fauzi
Department of Chemical Engineering and Sustainability, Kulliyyah of Engineering, International Islamic University Malaysia (IIUM), Kuala Lumpur, Malaysia
e-mail: jnoraini@iium.edu.my

D. N. Jimat (ed.), *Deep Eutectic Solvents*, Engineering Materials,
https://doi.org/10.1007/978-981-95-1912-5_1

the recovery and reuse of ILs are often tedious and time-consuming. As a greener alternative, deep eutectic solvents (DESs) have emerged. Unlike ILs, DESs are less toxic and more environmentally friendly. They are also biocompatible, biodegradable, and can be produced easily from relatively inexpensive and widely available materials.

While ILs are ionic compounds, DESs are generally non-ionic. They are created through the combination of hydrogen bond donors and acceptors, often in the presence of additional components such as ammonium or metal salts. The strong hydrogen bonding between these components results in a liquid solvent with a melting point lower than that of any individual substance. As a result, DESs remain effective at ambient or moderately elevated temperatures, simplifying their use in various applications.

DESs are easy to prepare; their synthesis typically involves mixing the desired components at moderate temperatures, with no need for complex purification steps. This simplicity reduces production costs, making DESs more suitable for large-scale industrial applications. They can be tailored using different donors and acceptors to yield solvents with varying properties, such as hydrophilicity or hydrophobicity. Consequently, DESs find use in pharmaceuticals, catalysis, separation processes, and even environmental remediation.

DESs are often considered greener than ILs and traditional solvents due to their biodegradability and lack of harmful vapor emissions. Their low toxicity and minimal environmental impact make them an ideal choice for industries looking for sustainable solutions. A lot of ongoing research on DESs has been conducted recently to explore new formulations and applications to meet the growing demand for green chemicals. It is expected that DESs will gain ground in the future, replacing ILs and other harmful solvents in applications such as drug production, where eco-friendly processes are increasingly required.

2 Discovery of Deep Eutectic Solvents (DES)

Deep eutectic solvents (DESs), discovered by Andrew Abbott and his coworkers in the early 2000s, marked a significant advancement in the field of green chemistry. Their development was driven by the need to find safer and more environmentally friendly alternatives to ionic liquids (ILs). Despite the versatility and widespread use of ILs across various industries, their drawbacks—such as high toxicity, environmental risks, and costly production processes—highlight the urgent need for better solvent options. DESs were introduced to overcome these limitations.

Abbott and his coworkers experimented with combining ammonium salts—among them choline chloride—with hydrogen bond donors (HBDs) such as urea, glycerol, and carboxylic acids, as later identified by Zhang and co-authors [18]. As a result, they produced liquids with low melting points due to the formation of strong hydrogen bonds between the components, even at ambient temperatures. These solvents proved to be effective under various conditions, offering stable surfaces for applications, and were thus suitable for industrial use, as reported by Dai and co-authors [3].

DESs gained popularity because their simple synthesis, non-toxicity during handling, and minimal environmental hazards. In contrast to ILs, which consist of a single cation and anion, DESs are formed through the combination of hydrogen bond donors and acceptors. This combination results in a non-toxic, biodegradable liquid that is less harmful to the environment compared to other solvents [16]. DESs can be simply prepared by mixing the components at moderately low temperatures, without the need for sophisticated filtration or any purification steps. Thus, they are cost-effective and appropriate solvents for large-scale industrial applications.

Thus, the emergence of DESs offers new horizons for greener chemicals and processes, with applications in various fields, including catalysis, chemical reactions, separation processes, and more. Their low molecular weight, combined with complete biodegradability, is an added advantage in applications where sustainability is the primary goal. The properties of DESs can be manipulated for specific industries by changing the composition of their components.

3 Deep Eutectic Solvent (DES)

Deep eutectic solvents (DESs) are a novel family of organic solvents first discovered by Abbott and co-authors [1]. DESs have been reported as an effective and sustainable technique for fractionating and further refining biomass components into various end-products [13]. These solvents are generated from eutectic mixtures of Lewis or Brønsted acids and bases that contain a variety of anionic and/or cationic species. In contrast, conventional ionic liquids consist of systems comprising only one type of discrete cation and anion. DESs are a subclass of ILs, by mixing two or more compounds to form a eutectic mixture that has lower melting points than the individual components. This mixture occurs from the intermolecular hydrogen bond between the hydrogen bond donor (HBD) and the hydrogen bond acceptor (HBA). Although the underlying principles of DESs and ionic liquids differ, they share many similar features and properties [16].

DESs are considered green solvents due to their low toxicity, non-volatility, and potential for recyclability and biodegradability, making them attractive for the sustainable processing of various biomaterials [15]. They can be easily synthesized from cheap bulk chemicals by simply mixing the components with heat or even at room temperature. These characteristics make DESs particularly appealing from the standpoint of green chemistry. In general, DESs are created through the combination of a quaternary ammonium salt with a metal salt or a hydrogen bond donor. In this context, the hydrogen bond donor refers to the species donating a proton, while the hydrogen bond acceptor denotes to the species accepting a proton [10].

DESs are classified into four categories based on the nature of the hydrogen bond acceptors and donors used [16], as presented in Fig. 1. As shown in Fig. 1, DESs are primarily categorized based on the type of hydrogen bond donor (HBD) used. There are five types of DESs, each represented by the general formula R^+A^-xB, where R^+ denotes the cationic core, which could be ammonium, sulfonium, or phosphonium. According to Ijardar and co-authors, A and B are defined as Lewis bases with a halide anion and Lewis's acid, respectively. The formation of a complex between x and the Lewis or Brønsted acid B occurs, where x represents the number of molecules interacting with the anion [9].

DES can be broadly categorised into four types based on the nature of the components involved:

1. Type I DES: These include a quaternary ammonium salt mixed with a metallic salt. The maximum common place instance is a mixture of choline chloride and zinc chloride. Type I DES are generally utilized in programs requiring the dissolution of steel oxides and other metal-based totally compounds.
2. Type II DES: These consist of a quaternary ammonium salt mixed with a steel chloride hydrate. An instance is a combination of choline chloride and aluminium chloride hydrate. These DES are frequently used in electroplating and different metallurgical methods.
3. Type III DES: These are fashioned via mixing a quaternary ammonium salt with a hydrogen bond donor which includes urea, glycerol, or ethylene glycol. Type III

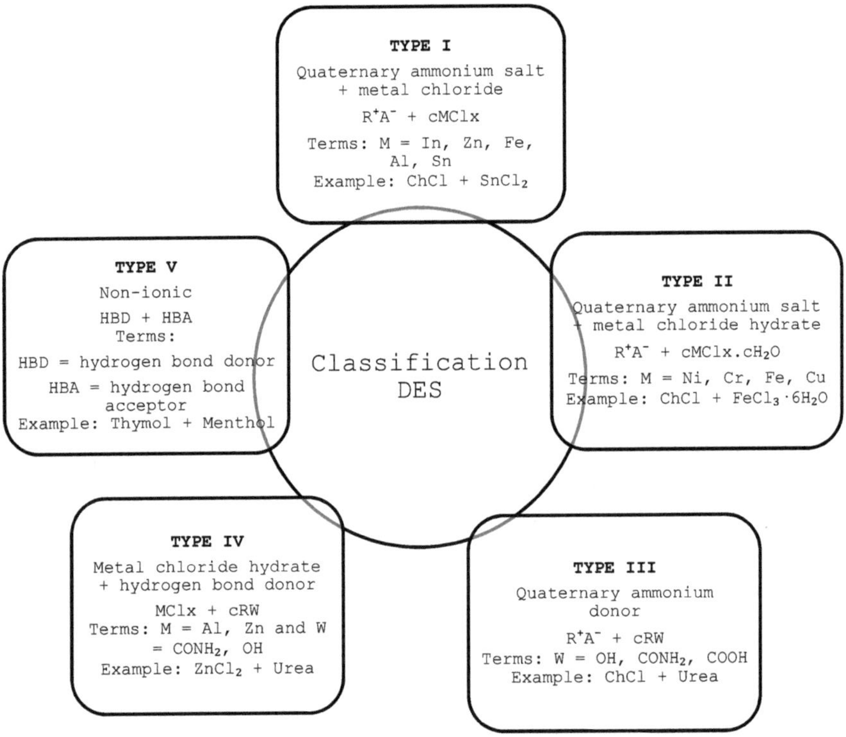

Fig. 1 Classification of DES based on different types of hydrogen bond donor [16]

DES are the maximum studied and widely used because of their low toxicity and biodegradable nature. A common example is a combination of choline chloride and urea.

4. Type IV DES: These consist of a steel salt and a hydrogen bond donor. An example is a combination of zinc chloride and ethylene glycol. These DES have capability packages in catalysis and different chemical techniques wherein steel-ligand interactions are crucial.

4 Properties of Deep Eutectic Solvents

4.1 Hydrophilic and Hydrophobic Nature

DESs can be either hydrophilic or hydrophobic, depending on the nature of the ingredients used during their formulation. The majority of DESs formulated to date are hydrophilic—that is, they mix well with water. This is due to components such as choline chloride and other similar ammonium compounds, which can form strong hydrogen bonds with water molecule [1, 16]. This water compatibility allows hydrophilic DESs to be used in specific processes such as biocatalysis, where enzymes require water to function effectively [4]. Previous studies have reported that enzymes perform excellently in these DESs, as they remain active in water-rich environments [18].

Hydrophobic DESs are less common but highly useful when working with non-polar (oil-like) substances. These DESs are typically prepared using ingredients such as fatty acids or long-chain alcohols that do not mix well with water. Hydrophobic DESs have also been considered non-toxic and green alternatives to traditional organic solvents. They are commonly used in the extraction of non-polar compounds—for example, plant extracts used in cosmetic or pharmaceutical formulations. DESs are highly flexible, meaning their components can be adjusted to make them either hydrophilic or hydrophobic, depending on the intended application [19].

4.2 Physicochemical Properties

The ease in DES preparation minimizes harmful volatile organic compound emissions that minimize hazards to human health and the environment [16]. DES also possess outstanding solvation ability, therefore, they are capable of dissolving metals, polymers, and chemicals among others, according to Zhang et al. [18]. Because of this reason, DES are considered suitable in such processes as the extraction of metals, chemical transformation, and pharmaceuticals among others according to Dai et al. [4]. DES can be tailored to desired viscosity based on the species used in their preparation, therefore, their versatility in a variety of industrial applications such as in catalysts or lubricants according to Maugeri and co-authors [11].

Another remarkable characteristic of DESs is their relatively low melting points, typically lower than those of the individual components used in their preparation. This property allows DESs to remain in the liquid state over a wide temperature range, enabling their application in various industrial processes—from very cold to very hot conditions. Researchers can also modify the formulation by adjusting the components to tailor the melting point to specific requirements, thereby broadening the range of potential applications. The melting point of DESs depends on the strength of hydrogen bonding and van der Waals forces between the hydrogen bond donor (HBD) and hydrogen bond acceptor (HBA). The interaction between the HBD and HBA during the eutectic phase is driven by changes in entropy and lattice energy, resulting in a significant reduction in melting point. Other factors influencing the melting point include the alkyl chain structure of the HBD and HBA. Both ionic and non-ionic hydrophobic DESs that contain long-chain fatty acid HBDs generally require higher temperatures to melt.

Most DESs have high viscosity due to the extensive hydrogen bonding between their components, which reduces the mobility of free species in the mixture [5]. The chemical nature of the individual components also affects the viscosity of DESs. This high viscosity makes large-scale application of DESs more challenging. However, the viscosity of DESs decreases with the addition of water. Moreover, the effect of water on viscosity is temperature-dependent. At lower temperatures, the viscosity of DESs decreases with increasing water content [17]. Hydrogen bond donors (HBDs), such as sugar derivatives or carboxylic acids, typically result in high-viscosity DESs due to the strong intermolecular hydrogen bonding network.

Knowing the density of DES provides valuable insights into the intermolecular interactions occurring within DES solvents. Most DESs are denser than water. The density of DESs is associated with the molecular characteristics of the hydrogen bond donor (HBD), steric effects such as the salt-to-HBD ratio, temperature, and the strength and extent of interactions between ions and HBDs [8]. The density of DESs increases when the HBD contains more hydroxyl groups or diacid groups, which results in the formation of more hydrogen bonds and a reduction in the free volume available [2]. Other studies have reported that the density of a eutectic mixture may vary depending on the composition and molar ratio of the hydrogen bond acceptor (HBA) to the HBD [2, 14]. Temperature is another factor that influences the density of DESs, as density decreases with increasing temperature.

Surface tension is one of the basic fluid properties that may influence the impact of deep eutectic solvents (DESs) in interfacial mass transfer processes. Both temperature and the intensity of intermolecular interactions in DESs also affect surface tension. The surface tension of most DESs is greater than that of conventional solvents. The strength of the interactions between the hydrogen bond donor (HBD) and hydrogen bond acceptor (HBA) affects the surface tension of DESs.

Both the hydrogen bond acceptor (HBA) and hydrogen bond donor (HBD) affect solvation interactions. Generally, DESs with high concentrations exhibit high viscosity. This is due to the presence of more functional groups and longer alkyl chains in DESs, which lead to stronger interactions between the components. For example, the viscosity β-coefficient of glucose increases at higher concentrations of DES, due to enhanced hydrogen bonding between glucose and DES. DESs become more viscous due to the presence of longer alkyl chains or fluorinated organic components, which contribute to increase van der Waals forces and hydrogen bonding. DESs, with their multiple solvation interactions, are more easily manipulated compared to other organic solvents. Their tunable hydrogen-bonding capability makes them useful for solubilization and stabilization processes. The different solvation interactions between solutes and DESs influence the physical properties of the solvents. Thus, optimum solvation interactions and full utilization of DESs can be achieved by selecting the most suitable DES.

5 Comparison of Ionic Liquid and DES

Some of the researchers consider that DESs is a subclass of ILs, however some of them agreed that they are independent from each other in nature [16]. Basically, there are two things that differentiate DESs from ILs, which are the methods of its formation and the nature of the starting materials. ILs is formed through the mixture of organic heterocyclic cations and organic or inorganic anions, while DESs is formed through various combination between hydrogen bond acceptors (HBAs) and hydrogen bond donors (HBDs). However, there is unlimited number of mixtures can be used to form both ILs and DESs. The mixture between these combinations that influence the properties of these substances to be interchangeable to meet the final applications [12].

Most of the components used to form DESs are natural compounds that are not hazardous to environment, easily accessible, sustainable and cheap compounds. The preparation of DESs also considered simple as it not required high technology as it can be prepared by vigorous stirring at room temperature or applied a moderate heat until the clear liquid of DESs is formed. While for ILs preparation, there are several steps needed including the use of various reagents such as organic solvent that most of it is a volatile compound which not environmentally friendly. Initially, to prepare ILs, the individual compound of ILs need to be diluted using organic solvent before it can be mixed together. Later, the mixture will undergo heating at the temperature between 25 to 100 °C under constant stirring. Once a clear and homogeneous liquid formed, sometime the solid residue needs to be separated before it can be used. ILs preparation required longer reaction time (up to 48 h) and sometime, ILs is required to be prepared through grinding and freeze drying in which it is a bit more complicated [6, 7, 12]. Figure 2 shows the methods used in ILs preparation.

Properties of ILs and DESs are differ due to different intermolecular interaction. ILs has a known properties due to ionic interaction that influence most of its properties. Whereas DESs behaviour is exhibit from the hydrogen bond interaction between its components. Both ILs and DESs have a low melting point, where most of ILs have a melting point below 100°C while DESs have a melting point below its individual component. Other properties such as density and viscosity for both ILs and DESs have a similar value, and it influenced by their composition. Therefore, the

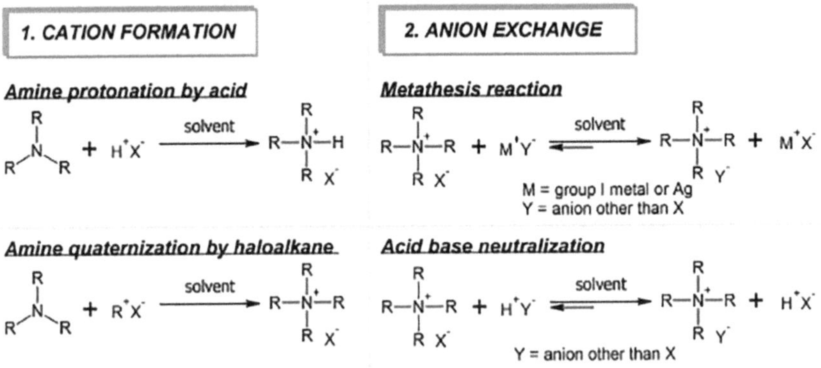

Fig. 2 Methods to prepare ILs (*adopted from* [12])

selection of combination for both constituents is important as it may influence the viscosity of the end liquid formed, and viscosity is the main drawback that hinder the application of these solvent in industry. Generally, ILs and DESs both have different chemical properties, but they have similar physical properties such as polarity, low vapour pressure and nonflammability [16]. In term of chemical properties, DESs is better in term of biodegradability and low toxicity which is due to the natural origin of its constituent. While for ILs, as the formation of this solvent require onium salts such as 1-alkylpyridinium and cholinium that is known to be high toxicity, thus ILs is considered not environmentally friendly as DESs. To date, many studies have been conducted on ionic liquids (ILs) and deep eutectic solvents (DESs), but more research has focused on ILs, as they were discovered earlier than DESs. Nevertheless, both solvents offer interesting prospects for further research, especially in analytical chemistry and industrial applications.

References

1. Abbott, A.P., Boothby, D., Capper, G., Davies, D.L., Rasheed, R.K.: Deep eutectic solvents formed between choline chloride and carboxylic acids: versatile alternatives to ionic liquids. J. Am. Chem. Soc. **126**(29), 9142–9147 (2004). https://doi.org/10.1021/ja048266j
2. Basaiahgari, A., Panda, S., Gardas, R.L.: Effect of ethylene, diethylene, and triethylene glycols and glycerol on the physicochemical properties and phase behavior of benzyltrimethyl and benzyltributylammonium chloride based deep eutectic solvents at 283.15–343.15 K. J. Chem. Eng. Data **63**(7), 2613–2627 (2018). https://doi.org/10.1021/acs.jced.8b00213
3. Choi, Y.H., van Spronsen, J., Dai, Y., Verberne, M., Hollmann, F., Arends, I.W.C.E., Witkamp, G.J., Verpoorte, R.: Are natural deep eutectic solvents the missing link in understanding cellular metabolism and physiology? Plant Physiol. **156**(4), 1701–1705 (2011). https://doi.org/10.1104/pp.111.178426
4. Dai, Y., van Spronsen, J., Witkamp, G.J., Verpoorte, R., Choi, Y.H.: Natural deep eutectic solvents as new potential media for green technology. Anal. Chim. Acta **766**, 61–68 (2013). https://doi.org/10.1016/j.aca.2012.12.019

5. de Oliveira Vigier, K., García-álvarez, J.: Deep eutectic and low-melting mixtures. In: Jérôme, F., Luque, R. (eds.) Bio-Based Solvents, 1st edn., pp. 83–114. John Wiley & Sons Ltd. (2017). https://doi.org/10.1002/9781119065357.ch4
6. Florindo, C., Oliveira, F.S., Rebelo, L.P.N., Fernandes, A.M., Marrucho, I.M.: Insights into the synthesis and properties of deep eutectic solvents based on cholinium chloride and carboxylic acids. ACS Sustain. Chem. Eng. **2**(10), 2416–2425 (2014). https://doi.org/10.1021/sc500439w
7. Gutiérrez, M.C., Ferrer, M.L., Mateo, C.R., Monte, F.D.: Freeze-drying of aqueous solutions of deep eutectic solvents: a suitable approach to deep eutectic suspensions of self-assembled structures. Langmuir **25**(10), 5509–5515 (2009). https://doi.org/10.1021/la900552b
8. Hansen, B.B., Spittle, S., Chen, B., Poe, D., Zhang, Y., Klein, J.M., Horton, A., Adhikari, L., Zelovich, T., Doherty, B.W., Gurkan, B., Maginn, E.J., Ragauskas, A., Dadmun, M., Zawodzinski, T.A., Baker, G.A., Tuckerman, M.E., Savinell, R.F., Sangoro, J.R.: Deep eutectic solvents: a review of fundamentals and applications. Chem. Rev. **121**(3), 1232–1285 (2021). https://doi.org/10.1021/acs.chemrev.0c00385
9. Ijardar, S.P., Singh, V., Gardas, R.L.: Revisiting the physicochemical properties and applications of deep eutectic solvents. Molecules **27**(4), 1–37 (2022). https://doi.org/10.3390/molecules27041368
10. Ma, Y., Xia, Q., Liu, Y., Chen, W., Liu, S., Wang, Q., Liu, Y., Li, J., Yu, H.: Production of nanocellulose using hydrated deep eutectic solvent combined with ultrasonic treatment. ACS Omega **4**(5), 8539–8547 (2019). https://doi.org/10.1021/acsomega.9b00519
11. Maugeri, Z., Leitner, W., Domínguez De María, P.: Chymotrypsin-catalyzed peptide synthesis in deep eutectic solvents. Eur. J. Org. Chem. **20**, 4223–4228 (2013). https://doi.org/10.1002/ejoc.201300448
12. Plotka-Wasylka, J., de la Guardia, M., Andruch, V., Vilkova, M.: Deep eutectic solvents vs. ionic liquids: similarities and differences. In: Current Developments in Biotechnology and Bioengineering: Deep Eutectic Solvents: Fundamentals and Emerging Applications, vol. 159, pp. 105–138 (2022). https://doi.org/10.1016/B978-0-323-99905-2.00007-8
13. Scelsi, E., Angelini, A., Pastore, C.: Deep eutectic solvents for the valorisation of lignocellulosic biomasses towards fine chemicals. Biomass (Switzerland) **1**(1), 29–59 (2021). https://doi.org/10.3390/biomass1010003
14. Shafie, M.H., Yusof, R., Gan, C.Y.: Synthesis of citric acid monohydrate-choline chloride based deep eutectic solvents (DES) and characterization of their physicochemical properties. J. Mol. Liq. **288** (2019). https://doi.org/10.1016/j.molliq.2019.111081
15. Sirviö, J.A., Visanko, M., Liimatainen, H.: Acidic deep eutectic solvents as hydrolytic media for cellulose nanocrystal production. Biomacromolecules **17**(9), 3025–3032 (2016). https://doi.org/10.1021/acs.biomac.6b00910
16. Smith, E.L., Abbott, A.P., Ryder, K.S.: Deep eutectic solvents (DESs) and their applications. Chem. Rev. **114**, 11060–11082 (2014)
17. Xu, H., Peng, J., Kong, Y., Liu, Y., Su, Z., Li, B., Song, X., Liu, S., Tian, W.: Key process parameters for deep eutectic solvents pretreatment of lignocellulosic biomass materials: a review. Biores. Technol. **310**(April), 123416 (2020). https://doi.org/10.1016/j.biortech.2020.123416
18. Zhang, Q., De Oliveira Vigier, K., Royer, S., Jérôme, F.: Deep eutectic solvents: syntheses, properties and applications. Chem. Soc. Rev. **41**(21), 7108–7146 (2012). https://doi.org/10.1039/c2cs35178a
19. Zhao, B.Y., Xu, P., Yang, F.X., Wu, H., Zong, M.H., Lou, W.Y.: Biocompatible deep eutectic solvents based on choline chloride: characterization and application to the extraction of rutin from Sophora japonica. ACS Sustain. Chem. Eng. **3**(11), 2746–2755 (2015). https://doi.org/10.1021/acssuschemeng.5b00619

Dissolution of Plastics Using DESs

Dzun Noraini Jimat, Nurul Sakinah Engliman, and Nurulnadiah Jasmin Mohamed Daud

Abstract Studies related to dissolution-based recycling using deep eutectic solvents (DESs) are still scarce. Therefore, this chapter covers the dissolution mechanism of polymer–solvent systems, particularly focusing on DES-based dissolution of PET and the subsequent processes after dissolution. PET recycling via DES-catalyzed glycolysis, the practicality of dissolution-based recycling for plastic waste, and the associated challenges are also discussed in this chapter. Suitable solvents must be used to ensure that only one polymer remains in the solid state at each step. It is also important to note that the selected solvents may vary depending on the plastic composition. Solvent-based recycling can serve as an alternative to mechanical and chemical recycling. In dissolution-based recycling, DESs act either as solvents or catalysts, providing nucleophilic sites for solvolysis and facilitating the development of synthetic processes that are more biocompatible and environmentally friendly compared to traditional solvolysis methods for plastic materials. Furthermore, their application enables mild depolymerization reaction conditions, reducing energy requirements without compromising performance.

1 Introduction

Massive accumulation of plastic wastes around the world has become a serious problem due to its endless consumption day by day. Since every object is made of this material, its durability, low cost of manufacture, adaptability, and difficulty of degradation led to its widespread use in a variety of applications. Around 380 million tons of plastic are produced annually worldwide, and by the end of 2050, that amount will have consistently surpassed 30 million tons [4, 7]. Additionally, the breakdown of these non-degradable wastes produces microplastics, which pose a serious threat to aquatic and marine ecosystems [21]. There are several efforts to recycle, reuse,

D. N. Jimat (✉) · N. S. Engliman · N. J. Mohamed Daud
Department of Chemical Engineering and Sustainability, Kulliyyah of Engineering, International Islamic University Malaysia (IIUM), Kuala Lumpur, Malaysia
e-mail: jnoraini@iium.edu.my

D. N. Jimat (ed.), *Deep Eutectic Solvents*, Engineering Materials,
https://doi.org/10.1007/978-981-95-1912-5_2

however just 7% of million tons of plastic wastes generated gets recycled annually as reported by Environmental Protection Agency [9].

One of the primary plastics that is frequently used to package goods like food, water, automobile components, textiles, and many more is polyethylene terephthalate, or PET. It shows that its widely consumption and demands in various manufacturing sectors. This is because of its exceptional qualities—such as resistance to heat, mechanical stress, and chemicals, as well as lightweight, stability, and durability [9]. However, its ongoing production and use have resulted in environmental issues around the world, which must be addressed. PET, a kind of polyester which is a synthetic thermoplastic that can be either semi-crystalline (white and opaque) or amorphous (transparent) [3].

Currently, pyrolysis and gasification are often used to process these plastic wastes which need to operate at high temperatures (500–1000 °C) and subsequent upgrading to process the complex mixture of products. Chemical recycling plastic wastes via oxidation degradation using nitric acid at high temperature is a relatively harsh condition and requires high resistance equipment. Thus, an innovative and sustainable way to process these plastic wastes and convert it to value added products is highly vital. Deep eutectic solvents (DESs) represent a new class of green and sustainable solvents with negligible toxicity, biodegradability, ease of synthesis which potential to be used in dissolution of polymer. Dissolution-based polymer method is one of recycling approaches that able to produce highly free-impurities polymers, dissolve specific polymer even from varies plastic component and commonly it is eco-friendly compared to chemical recycling [2]. This approach comprises three steps which firstly, dissolving the polymer using appropriate solvent or solvent mixture at certain temperatures following by separating the non-dissolved solids and the solvated polymers in liquid phase via filtration or centrifugation. Lastly precipitating the solvated polymers using antisolvent that promotes its insolubility indirectly removing the used solvent.

Studies related to dissolution-based recycling using deep eutectic solvent (DES) are still scarce. Hence, this chapter covers the dissolution mechanism of the polyethylene terephthalate (PET) plastic wastes into value added products such as virgin resins using deep eutectic solvent (DES), suitability of DES mixtures in dissolution of plastic resins. Additionally, the challenges of the DES mixtures and their practicality for use in recycling plastic wastes compared to other plastic wastes recycling strategies are discussed too.

2 Dissolution Mechanism of Polymers–Solvent

The physico-chemical characteristics of polymers, including their types, sizes, and molecular weights, as well as the parameter conditions, including temperature, concentrations, dissolving time, and solvent selection, all could affect the effectiveness of dissolution-based polymer strategy [21]. The mechanisms of dissolution of polymer are very complicated as it relates with chemical interaction between solute, solvent and solid polymer matrix [14]. Its mass transfer mechanism is influenced by the permeability of solid polymer matrix. During the dissolution process, the solvent penetrates the permeable matrix indirectly causes the presence of soluble substances in solvent at a specific rate and this phenomenon governed by the diffusivity [14]. Thus, it shows that the permeability relies on both solubility and diffusivity. Equation below describes the relationship of this mechanism [14]:

$$P = D \cdot S$$

where P = the permeability coefficient (mol/ms Pa), D = the diffusion coefficient (m^2/s) and S = the solubility coefficient (mol/m^3 Pa).

Understanding the behaviour of DES is utmost important as it determines its role in any specific application. Hydrophobic DESs with lesser viscosities (<20 mPa) are favourable compared to conventional organic solvents such as benzene, octane, heptane and others [20]. DESs viscosities can be tuneable by manipulating its temperature, molar ratio and composition of HBA/HBD. Moreover, a moderate viscosity is essential to smoothen agitation and centrifugation. Perhaps, DESs can be an effective solvent to be used for dissolution of plastic wastes due to its wider range of viscosities compared to conventional volatile solvents. Additionally, their densities are greater than water except for certain hydrophobic DESs exhibit lower density [19]. This property could affect the strength and solubility of the solvent and might result in the formation of polymorphs. The selection of the best solvents is mostly based on a few key guidelines [21]. First, high polar polymers dissolve in high polar solvents, while low polar polymers dissolve in low polar solvents, in accordance with the solubility rule "like dissolves like." The polymer dissolves more readily in the solvent if the solvent and polymer have similar or equal solubility parameters; second, the polymer dissolves more readily in the solvent when the solvent–polymer interaction outweighs the polymer's cohesiveness; and third, the polymer dissolves more easily in the solvent if the solvent and polymer have similar or equal solubility parameters (δ_D, δ_P, δ_H).

δ_D: interaction dispersion forces

δ_P permanent dipole-permanent dipole forces

δ_H hydrogen bonding.

Selection of appropriate solvent can be made based on Hansen solubility parameter (HSP) values which correlate with three solubility parameters; δ_D, δ_P, and δ_H [5]. These solubility parameters depend on temperature and molecular size and shape. The solubility limit of polymer in solvent can be measured using the relative energy difference (RED). RED is known as the ratio of R_a (distance from a given solvent point to the centre of the sphere) to R_0 (the radius of sphere).

$$\mathrm{RED} = \frac{R_a}{R_0}$$

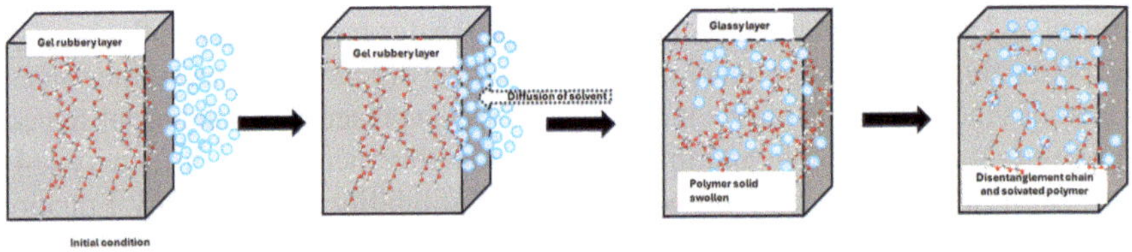

Fig. 1 The solvated polymer molecules that are distributed throughout the solution and the polymer chains separate from the surface of the swelling polymer

RED < 1	RED ≈ 1	RED > 1
Soluble and high affinity; strong solvent	Partially soluble; boundary condition	Non-soluble and low affinity; weak solvent

where $(R_a)^2 = 4(\delta_{D1} - \delta_{D2})^2 + (\delta_{P1} - \delta_{P2})^2 + (\delta_{H1} - \delta_{H2})^2$

Solubility parameter "distance" R_a between a polymer (subscript 1) and the solvent (subscript 2).

After the polymer successfully diffuses into the appropriate solvent, it may begin to swell. It now resembles a gel-rubbery swelling layer that has created two interfaces: one between the gel layer and the glassy polymer, and another between the gel layer and solvent. After a while, as seen in Fig. 1, the solvated polymer molecules are distributed throughout the solution and the polymer chains separate from the surface of the swelling polymer.

Another significant phenomenon in polymer solutions is diffusivity whereas it is connected to the viscosity. Diffusion of the solvent is related to the physical properties of the polymer and with the interactions between the polymer and the solvent. A number of variables such as temperature, molar mass, concentration, and polymer morphology, including crystallinity, crosslinking, and orientation, affect the diffusion rate, or the number of molecules that diffuse over time. The Deborah number (De) can be used to analyse the behaviour observed in solvent diffusion processes [13]. This number is expressed as the ratio between the characteristics time of the fluid (τ) and the time of the deformation process (t_p), De $= \tau/t_p$ and signifies the degree of viscoelasticity [6]. A Deborah number with zero value correspond to a viscous fluid, while an infinite Deborah number indicates the elastic behavior [6, 13].

2.1 *DESs-Based Dissolution of PET*

Currently studies related to PET recycling through DESs-based dissolution approaches are still scarce. Pestana [12] evaluated the solubility coefficient of DES in dissolution of PET by determining Hansen's solubility parameters (HSP) coupled with the Stefanis Panayiotou group contribution [12]. This study found that the value of relative energy difference (RED) of DES mixtures; thymol:carvacrol (1:1) was 0.746 which is less than one compared to other DES mixtures. This dissolution process was conducted at 150 °C for 7 min. It showed that the PET dissolution process with DES mixtures can be carried out under mild conditions. The use of thymol:carvacrol serves stronger hydrogen bonding between DES and PET due to availability of the OH– groups [12]. Since thymol/carvacrol is a hydrophobic DES mixture consequently, the dissolved PET can be recovered by using low cost antisolvent such as water and ethanol.

2.2 Subsequent Process After Dissolution

Precipitation process is a common step after the dissolution process of the polymeric constituents via cooling or evaporating the solvent or inclusion of nonsolvent with different polarity and solubility [14]. Selection of appropriate antisolvent is highly important to ensure the polymeric constituents are properly precipitated. Principally, solvents are classified as polar and apolar solvents for example polymers with polar groups like PET can be dissolved in polar solvents such as trifluoro-ethanol and hexafluoropropane [1, 14]. In some cases, mixed solvents are used (i.e. toluene/chloroform) to dissolve a PE matrix at 105 °C [16]. Precipitation of polymers can be also carried out by cooling the solution instead of adding a nonsolvent [14]. This method simplifies the cleaning process and might decrease the cost considerably. Selection of a suitable non-solvent with different polarity is also important for precipitation of polymer constituents to ensure it achieves a desirable amount of precipitated polymer constituents, avoiding excessive amounts of nonsolvent which later is not sustainable and costly.

Walker and co-authors discovered the solvent-targeted recovery and precipitation (STRAP) methods which selectively dissolve a single polymer layer of multilayer polymers in a solvent system [15]. They reported that the specific solvents capable to dissolve each polymer of main multilayer polymer which consist of polyethylene, ethylene vinyl alcohol, and polyethylene terephthalate (PE, PET and EVOH). After filtering the polymer solution, it is precipitated by cooling and/or by adding the antisolvent. These processes are repeated until the rest of undissolved layers are recovered individually with changing the combination of solvent/antisolvent. They demonstrated this unique strategy, STRAP through guided approach which are Hansen solubility parameters (HSP), molecular dynamics (MD) simulations and combinations of quantum chemical and statistical mechanical approach known as the conductor-like screening model for realistic solvents (COSMO-RS) [15]. They showed that three polymers PE, PET and EVOH are recovered separately from the multilayer film and their quality is competitive with the virgin polymers through the STRAP approach [15].

3 PET-Recycling Via DES-Catalysed Glycolysis

Conventionally, depolymerization of PET via glycolysis is catalysed by metal derivatives, zeolites or ionic liquid and carried out at high temperature (>190 °C) and lengthy time (up to 480 min) [10]. In the absence of a catalyst, the glycolysis reaction rate might be low and hinder the completion of PET conversion into value-added products such as bis(hydroxyalkyl) terephthalate (BHET). Furthermore, its implementation in industry scale is hindered due to insignificant of the targeted yield. Thus, it is necessary to use DESs as solvents and catalysts which provide nucleophilic sites for solvolysis and facilitate the development of synthetic processes that are biocompatible and/or eco-friendly compared to traditional solvolysis of plastic materials [11]. Additionally, its application provides mild depolymerization reaction conditions, lowering energy requirements without compromising its performance.

A high PET conversion and great selectivity of bis(hydroxyalkyl) terephthalate (BHET) monomer were successfully achieved via glycolysis which catalysed with urea/metal salt DESs [17]. The used of catalyst (n(urea)/n($ZnCl_2$) (4:1) in this PET glycolysis which was conducted at 170 °C for 30 min was more effective compared to single catalyst, such as only urea, $ZnCl_2$, $ZnAc\cdot 2H_2O$ or $MnAc_2\cdot 4H_2O$ (Scheme 1). This result showed that a high catalytic activity of DES causes the synergetic catalysis of H-bonds and coordination bonds formed between this catalyst and ethylene glycol (EG) [17]. The same results were demonstrated but in this study, 1,3-dimethylurea (1,3-DMU)/$Zn(OAc)_2$ DES was used and the glycolysis occurred at 190 °C with duration of 20 min to complete PET conversion [8] (Scheme 2). Other studies demonstrated the PET catalytic glycolysis but betaine/$Zn(OAc_2$ DES was used as a catalyst [8] (Scheme 3). Zhu et al. reported that 90% of BHET were obtained from glycolysis of PET which catalysed by 1,5,7-triazabicyclo [4.4.0]dec-5enium acetate [HTBD(OAc)]/$Zn(OAc)_2$ DES [22]. Even after seven cycles, this kind of DES catalyst maintained its catalytic activity (Scheme 4). Conversion of PET to BHET with 83.2% yield was also reported by Wang et al. [18] via glycolysis with catalyst, acetamide/Znl_2 supported on zeolitic imidazole framework type 8 (ZIF-8) (Scheme 5) [18]. This study also employed the recovered-filtered catalyst for six times in PET glycolysis with minimal loss of catalytic activity. These different 5 schemes of PET catalytic glycolysis are summarized in Fig. 2.

4 Practicability of Dissolution-Based Recycling for Plastic Wastes

Several factors should be considered during the practical application, such as the dissolution ability of the solvent, viscosity, toxicity, time consumed for the dissolution, forms of the obtained polymer, and cost [14].

Viscosity of the polymer solution is one of the variables that may influence the effectiveness of solvent-based recycling of plastic wastes and the subsequent process

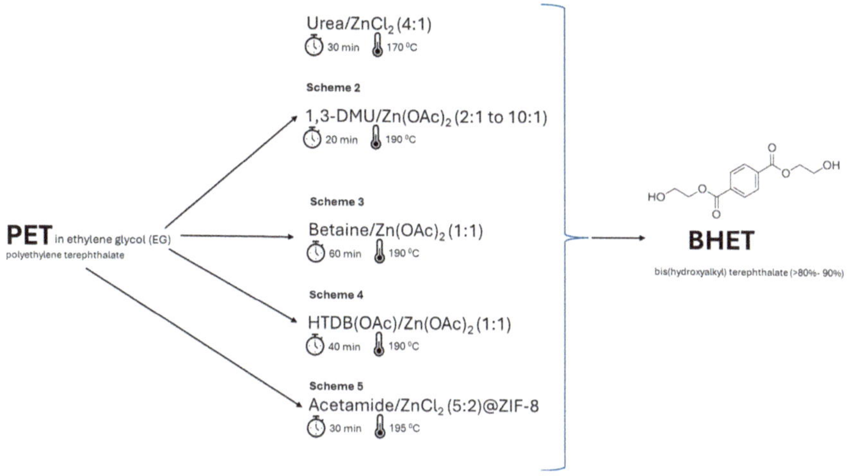

Fig. 2 Five schemes of PET catalytic glycolysis

to recover the solvated polymer. The viscosity of polymer solutions is dependent on concentration and molar mass of polymer solutions, diffusivity, temperature and pressure as well as polymer/solvent interactions [6]. The higher concentration of polymer solutions, the higher its viscosity because of the existence of polymer entanglements [6]. For the constant polymer concentration with average high molar mass possessing high viscosity may lead to low shear rates. However, at high shear rates, the viscosity curve on log–log scale results in a straight line regardless of the average molar mass. Polymer configuration is also affecting the viscoelastic properties as the higher degree of polymer chain branching, the higher its entanglements and consequently increases its viscoelastic properties. A high temperature may cause decrement of viscosity of the polymer solution and melts. It is because the high temperature promotes the movement of polymer molecules because of less solvent viscosity and inter-chain contacts and enhances the polymer solubility. High temperature leads to a short relaxation time whereas an increase in pressure increases the relaxation time. When the stress is ceased and the material turns back to its initial structure throughout a characteristic time, this time is called relaxation time. This time can be applied as an explicit extent of the relative value of viscosity and elasticity. Fluid with a higher relaxation time than other fluids indicates that it possesses a lesser elasticity compared to other fluids [6].

5 Challenges

Even though dissolution–precipitation method is considered minimal cost and able to remove a whole range of additives but selection of appropriate solvent/nonsolvent mixture is crucial specifically for polymers that are challenging to dissolve [6].

Occasionally, high temperatures are required however it may raise some concerns related to viscosity, corrosion, solvent concentrations which later lead to safety issues.

Several mass transfer stages are involved in the dissolution process, which could lengthen the process and ultimately make it more difficult to adopt on an industrial scale where cost and processing time are concerns. Thus, it needs fast processes especially when involved in high volumes at large scale to ensure an effective dissolution process. Furthermore, prior processes such as removal of contaminants/ink or mechanical shredding before dissolution could accelerate the dissolution kinetics of the polymer wastes.

Suitable solvents must be used to leave one polymer in solid state in each step. It is also noted that the selected solvents may vary depending on the plastic compositions. Toxic solvents even with ideal dissolution ability were not recommended to establish the recycling method. A moderate viscosity is required to facilitate stirring and centrifugation. Recycling methods with a long processing time were not practical for industrialization. Furthermore, recovery of the dissolved polymers could be challenging specifically when dealing with viscous polymers. Highly diluted polymers, however, raise issues, such as requirement to recover a significant volume of the utilized solvent and the requirement for a bulk quantity of antisolvent to precipitate the dissolved polymers.

However, solvent-based recycling can be a viable alternative to mechanical and chemical recycling. It can be successfully employed commercially if a proper solvent management system is implemented, along with a clear understanding of the complex rheological behavior of the polymer with the chosen solvent, as many variables are involved (e.g., concentration, solvent type, temperature, pressure, and average molar mass). Other factors that need to be considered when selecting a solvent include miscibility, boiling point, cost, toxicity, operability, and environmental impact.

References

1. Braun, D. (2013). Simple Methods for Identification of Plastics, 5th edn. Hanser
2. Chaudhari, U.S., Kulas, D.G., Peralta, A., Hossain, T., Johnson, A.T., Hartley, D.S., Handler, R.M., Reck, B.K., Thompson, V.S., Watkins, D.W., Shonnard, D.R.: Solvent based dissolution-precipitation of waste polyethylene terephthalate: economic and environmental performance metrics. RSC Sustain. **1**(7), 1849–1860 (2023). https://doi.org/10.1039/d3su00231d
3. Dhaka, V., Singh, S., Anil, A.G., Sunil Kumar Naik, T.S., Garg, S., Samuel, J., Kumar, M., Ramamurthy, P.C., Singh, J.: Occurrence, toxicity and remediation of polyethylene terephthalate plastics. A review. Environ. Chem. Lett. **20**(3), 1777–1800 (2022). https://doi.org/10.1007/s10311-021-01384-8
4. Geyer, R., Jambeck, J.R., Law, K.L.: Production, use, and fate of all plastics ever made. Sci. Adv. **3**(7), 25–29 (2017). https://doi.org/10.1126/sciadv.1700782
5. Hansen, C.M., Just, L.: Prediction of environmental stress cracking in plastics with Hansen solubility parameters. Ind. Eng. Chem. Res. **40**(1), 21–25 (2001). https://doi.org/10.1021/ie9904955
6. Kol, R., De Somer, T., D'hooge, D.R., Knappich, F., Ragaert, K., Achilias, D.S., De Meester, S.: State-of-the-art quantification of polymer solution viscosity for plastic waste recycling. ChemSusChem **14**(19), 4071–4102 (2021). https://doi.org/10.1002/cssc.202100876

7. Li, H., Aguirre-Villegas, H.A., Allen, R.D., Bai, X., Benson, C.H., Beckham, G.T., Bradshaw, S.L., Brown, J.L., Brown, R.C., Cecon, V.S., Curley, J.B., Curtzwiler, G.W., Dong, S., Gaddameedi, S., García, J.E., Hermans, I., Kim, M.S., Ma, J., Mark, L.O., et al.: Expanding plastics recycling technologies: chemical aspects, technology status and challenges. Green Chem. **24**(23), 8899–9002 (2022). https://doi.org/10.1039/d2gc02588d
8. Liu, B., Fu, W., Lu, X., Zhou, Q., Zhang, S.: Lewis acid-base synergistic catalysis for polyethylene terephthalate degradation by 1,3-dimethylurea/Zn(OAc)$_2$ deep eutectic solvent. ACS Sustain. Chem. Eng. **7**(3), 3292–3300 (2019). https://doi.org/10.1021/acssuschemeng.8b05324
9. Naderi Kalali, E., Lotfian, S., Entezar Shabestari, M., Khayatzadeh, S., Zhao, C., Yazdani Nezhad, H.: A critical review of the current progress of plastic waste recycling technology in structural materials. Curr. Opin. Green Sustain. Chem. **40**, 1–7 (2023). https://doi.org/10.1016/j.cogsc.2023.100763
10. Padhan, R.K., Sreeram, A.: Chemical depolymerization of PET bottles via combined chemolysis methods. In: Recycling of Polyethylene Terephthalate Bottles, pp. 135–147 (2019). https://doi.org/10.1016/b978-0-12-811361-5.00007-9
11. Paparella, A.N., Perrone, S., Salomone, A., Messa, F., Cicco, L., Capriati, V., Perna, F.M., Vitale, P.: Use of deep eutectic solvents in plastic depolymerization. Catalysts **13**(7) (2023). https://doi.org/10.3390/catal13071035
12. Pestana, S.C.: Development of Green Solvents for Depolymerization of Polyesters (2020)
13. Sochi, T.: Flow of non-Newtonian fluids in porous media. J. Polym. Sci., Part B: Polym. Phys. **48**(23), 2437–2767 (2010). https://doi.org/10.1002/polb.22144
14. Ügdüler, S., Van Geem, K.M., Roosen, M., Delbeke, E.I.P., De Meester, S.: Challenges and opportunities of solvent-based additive extraction methods for plastic recycling. Waste Manag. **104**, 148–182 (2020). https://doi.org/10.1016/j.wasman.2020.01.003
15. Walker, T.W., Frelka, N., Shen, Z., Chew, A.K., Banick, J., Grey, S., Kim, M.S., Dumesic, J.A., Van Lehn, R.C., Huber, G.W.: Recycling of multilayer plastic packaging materials by solvent-targeted recovery and precipitation. Sci. Adv. **6**(47), 1–9 (2020). https://doi.org/10.1126/sciadv.aba7599
16. Wang, F.C.Y.: Polymer additive analysis by pyrolysis-gas chromatography: III lubricants. J. Chromatogr. A **891**, 313–324 (2000). https://doi.org/10.1016/S0021-9673(00)00346-0
17. Wang, Q., Geng, Y., Lu, X., Zhang, S.: First-row transition metal-containing ionic liquids as highly active catalysts for the glycolysis of poly(ethylene terephthalate) (PET). ACS Sustain. Chem. Eng. **3**(2), 340–348 (2015). https://doi.org/10.1021/sc5007522
18. Wang, R., Wang, T., Yu, G., Chen, X.: A new class of catalysts for the glycolysis of PET: deep eutectic solvent@ZIF-8 composite. Polym. Degrad. Stab. **183**, 109463 (2021). https://doi.org/10.1016/j.polymdegradstab.2020.109463
19. Weerasinghe, U.A., Wu, T., Chee, P.L., Yew, P.Y.M., Lee, H.K., Loh, X.J., Dan, K.: Deep eutectic solvents towards green polymeric materials. Green Chem. **26**(15), 8497–8527 (2024). https://doi.org/10.1039/d4gc00532e
20. Zainal-Abidin, M.H., Hayyan, M., Wong, W.F.: Hydrophobic deep eutectic solvents: current progress and future directions. J. Ind. Eng. Chem. **97**, 142–162 (2021). https://doi.org/10.1016/j.jiec.2021.03.011
21. Zhao, Y.B., Lv, X.D., Ni, H.G.: Solvent-based separation and recycling of waste plastics: a review. Chemosphere **209**, 707–720 (2018). https://doi.org/10.1016/j.chemosphere.2018.06.095
22. Zhu, C., Fan, C., Hao, Z., Jiang, W., Zhang, L., Zeng, G., Sun, P., Zhang, Q.: Molecular mechanism of waste polyethylene terephthalate recycling by the 1,5,7-triazabicyclo[4.4.0]decium acetate/zinc acetate deep eutectic solvent: the crucial role of 1,5,7-triazabicyclo[4.4.0]decium cation. Appl. Catal. A: Gen. **641**, 118681 (2022). https://doi.org/10.1016/j.apcata.2022.118681

Green Solvent-Based Approaches for Recycling Spent Lithium Batteries

Dzun Noraini Jimat, Muhammad Hasya Hasnuddin, and Sarina Sulaiman

Abstract This chapter covers common recycling methods for lithium-ion batteries (LiBs), including DES-assisted recycling of cathode active materials, binder dissolution and separation of electrode materials from spent LiBs using DESs, and the recovery and regeneration of anode graphite in DESs. The rapid and large-scale manufacturing of advanced portable electronic devices and electric vehicles (EVs) has led to a substantial increase in the production of high-capacity, rechargeable lithium-ion batteries. Improper disposal of spent LiBs can pose serious environmental and health risks, as they contain hazardous materials such as lead (Pb), cadmium (Cd), and lithium (Li). Therefore, it is crucial to develop sustainable recycling approaches to address the disposal issues of spent LiBs. The use of DESs allows for the efficient dissolution of lithium, cobalt, nickel, and other transition metals from spent cathode materials, often under milder conditions compared to traditional methods. However, further studies are needed before industrial-scale implementation to minimize the risk of lower extraction yields and to ensure the effective recovery of valuable materials.

1 Introduction

The vast manufacturing and advanced portable electronic devices and electric vehicles (EVs) caused the massive production of rechargeable and high-capacity lithium-ion batteries (LiBs). If this trend continues it may cause huge accumulation of spent LiBs and improper battery disposal may cause a vital impact to environment and human health as it is composed of hazardous materials such as Pb, Cd and Li. Thus, it is crucial to develop a sustainable recycling approach to overcome the disposal issues of the spent LiBs. The implementation of the recycling strategy must be effective, less-energy consumption, safe, cost-effective and eco-friendly to ensure successfully recovering the value-added materials such as Li, Co and Ni. Pyrometallurgy and hydrometallurgy are the common LiBs recycling strategies. A former

D. N. Jimat (✉) · M. H. Hasnuddin · S. Sulaiman
Department of Chemical Engineering and Sustainability, Kulliyyah of Engineering, International Islamic University Malaysia (IIUM), Kuala Lumpur, Malaysia
e-mail: jnoraini@iium.edu.my

D. N. Jimat (ed.), *Deep Eutectic Solvents*, Engineering Materials,
https://doi.org/10.1007/978-981-95-1912-5_3

strategy requires high energy as it needs to operate at high-temperature and only a few selected metals can be obtained in alloy form. The latter strategy uses acids and extractants to leach metals followed by a precipitation process of highly purified precursors for new cathode materials. This approach is less-energy needed with reduced CO_2 emissions. However, it is not environmentally friendly and harmful to the operators.

Generally, there are five different types of lithium-ion batteries (LiBs) which are lithium cobalt oxide (LCO), lithium iron phosphate (LFP), lithium manganese oxide (LMO), lithium nickel cobalt aluminium oxide (NCA) and lithium nickel manganese cobalt oxide (NMC). LiBs composed of negative and positive electrodes, electrolytes and a separator coiled in varying shapes and sizes [5, 11–14]. Presently, manufacturer is producing LiBs with higher specific energy, greater coulombic efficiency, lower self-discharge phenomenon, and many reagents in different versions of electrodes to accommodate widely different applications. Consequently, it may raise difficulty in recycling the spent LiBs. The emerging of green solvents, deep eutectic solvents (DESs) can be used in leaching valuable metals from the spent LiBs as an alternative to conventional solvents.

Deep eutectic solvents (DESs) have appeared as a promising alternative for the treatment of lithium-ion battery waste. DESs are a class of green solvents formed by the interaction of hydrogen bond donors (HBDs) and acceptors (HBAs), typically resulting in a eutectic mixture with a melting point significantly lower than that of its individual components. The unique properties of DESs, such as low volatility, non-flammability, high thermal and chemical stability, and tunable solubility, make them highly attractive for numerous applications, including metal recovery and recycling. Moreover, DESs are often biodegradable, cost-effective, and relatively easy to prepare, aligning well with the principles of green chemistry [11].

In the context of lithium-ion battery recycling, DESs have proven remarkable potential for the selective leaching and recovery of valuable metals. The use of DESs allows for the efficient dissolution of lithium, cobalt, nickel, and other transition metals from spent cathode materials, often under milder conditions compared to traditional methods. Furthermore, the ability to tailor the composition of DESs by selecting specific HBAs and HBDs enables the optimization of their chemical properties for targeted applications. For instance, choline chloride-based DESs combined with organic acids or urea have shown excellent performance in recovering metals from LIB wastes. Such systems not only decrease the environmental impact but also improve the economic feasibility of the recycling process by minimizing the need for additional purification steps [1, 9].

Pyrometallurgical recycling steps

Advantage

- appropriate to different types of LiBs cell systems, includes different cathode chemistries (e.g., lithium cobalt oxide, lithium iron phosphate, lithium nickel manganese cobalt oxide) and different electrolytes.

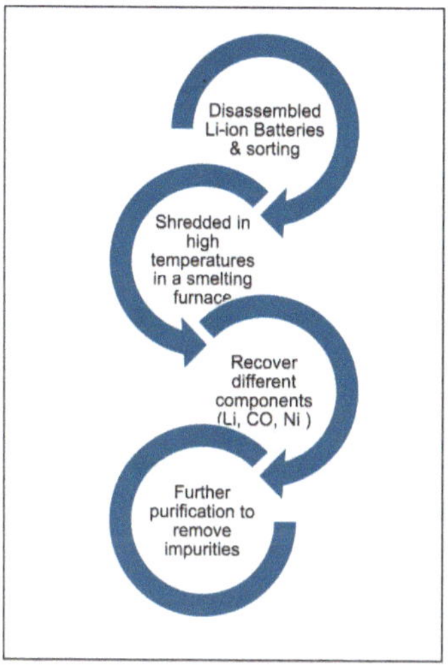

Disadvantages

- Excessive energy usage due to the high temperatures required for smelting, which may increase processing costs.
- emit pollutants such as greenhouse gases and lethal fumes.
- Lost or degraded of some materials during the smelting process
- Expose safety risks to labor and need strict safety measures.

Fig. 1 Pyrometallurgical recycling steps

2 Common Recycling Methods for Lithium-Ion Batteries (LiBs)

Implementing effective battery recycling strategies is crucial to ensure the implemented technology sustainable, eco-friendly, non-hazardous to human and conserve valuable elements as well as reduced the used of landfill space. Three common recycling approaches for lithium ions batteries (LiBs) and its drawbacks are summarized below (Figs. 1, 2 and 3).

3 DES-Assisted of Recycling Cathode Active Materials

Conventionally, leaching of valuable materials from spent lithium-ion batteries (LiBs) can be conducted using acids (such as sulfuric acid or hydrochloric acid) or alkaline (such as sodium hydroxide or potassium hydroxide). Acids dissolve metals more selectively and effectively compare to alkaline as the separation and recovery processes for certain metals such as cobalt (Co), are easier. However, this approach must be handled carefully, as it is harsh and generates hazardous byproducts that negatively impact human health and environment.

Although metal dissolution using alkaline is milder compared to acids, it still requires an additional neutralization step. Thus, there is a necessity for alternative solvents to overcome these challenges. Deep eutectic solvents (DESs) are promising

Hydrometallurgical recycling steps

Description

- Selection of leaching agent rely on recovery of the targeted metal and composition of the battery materials.
- An optimal metal dissolution and low dissolution of undesired materials can be achieved by controlling factors such as pH, temperature and agitation.
- Selectivity of leaching solution and enhancement of metal recovery efficiency can be attained by adding the chelating agents.
- Leach solution (pregnant liquor) can be separated from the solid residues (leach residue) using filtration, sedimentation or centrifugation.
- Then, the precipitation, solvent extraction or electrodeposition is subsequent process to recover the metal dissolved in the leach solution.

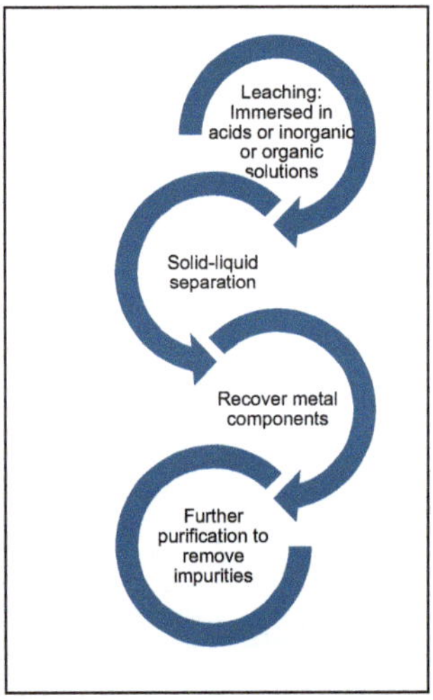

Disadvantages

- This recycling process produces better material purity compared to pyrometallurgical recycling method.
- However, this method requires substantial improvements to minimize the environmental impact as well as to simplify its process

Fig. 2 Hydrometallurgical recycling steps

candidates for recycling various types of LiBs due to their tunable solvation properties, which can be adjusted for specific metals by manipulating their compositions. This allows DESs to dissolve various metals with high selectivity and efficient metal recovery. Furthermore, DESs consume less energy and incur lower operating costs, as they can be used at lower temperature, generate less wastes, and are easier to manage.

A pure lithium (Li) element was successfully extracted from spent $LiFePO_4$ batteries using a combination of choline chloride (ChCl) and ethylene glycol (EG) with presence of ozone [12]. This study demonstrated that the combined effects of DES and ozone through acidification, replacement and oxidation reactions influenced the phase transition of Li and Fe. Consequently, this facilitated the active and selective extraction of Li while concurrently separating Fe in situ as $FePO_4$ [12]. Table 1 summarizes the leaching of spent LiBs using various DES along with their respective conditions and effectiveness.

The effectiveness of leaching LiBs components can be enhanced by using the appropriate types of deep eutectic solvents (DESs) under optimal leaching conditions. Physicochemical characteristics such as viscosity and acidity or basicity, may influence the leaching efficiency of LiBs components. Studies have shown that DES with

Mechanical recycling steps

Description of processes

- Grind or mill the shredded material to reduce its size and attain constant particle sizes.
- Sorting: separate different components according to size, density, and magnetic properties.
- Metallic foils (usually Al and Cu), which are used as current collectors, can be removed from electrode materials via sieving, gravity separation, or magnetic separation.
- Li, Co, Ni, and other metal-based electrode materials are separated from binders and other constituents via mechanical sieving, air classification, or flotation.

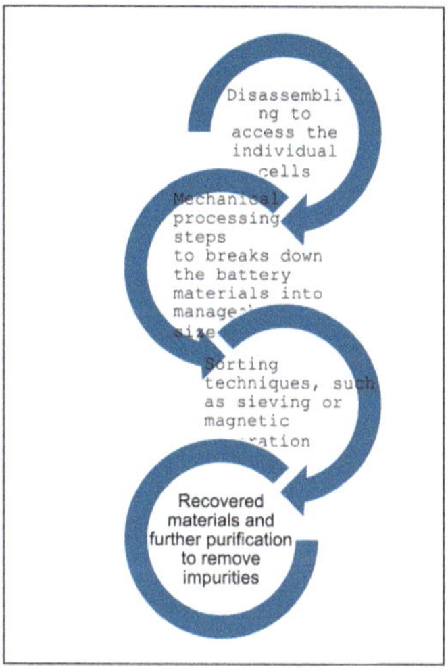

Disadvantages

- Material purity might not attain same degree as other recycling approach; challenging process if rely on mechanical separation alone
- Consume high energy input, specifically during shredding and size reduction stages
- exposure to hazardous compounds present in the battery cells
- can influence the overall environmental footprint of the recycling process

Fig. 3 Mechanical recycling steps

low viscosity are preferred and those with stronger acidity result in higher leaching efficiency for various elements. The leaching rate can also be slightly increased by the presence of small amount of water as it may improve transport properties under high-temperature and extended time of leaching. Additionally, the leaching process can be enhanced by assistance of other techniques such as ultrasound, microwave and inclusion of external agents such as aluminium (Al) and copper (Cu).

4 Binder Dissolution and Separation of Electrode Materials of Spent LiBs Using DESs

Generally, cathode materials adhere tightly to aluminium (Al) foil, which serves as the current collector in different kinds of lithium-ion batteries (LiBs), using the organic binder. Binder dissolution via DESs is a promising approach for recycling spent LiBs offering advantages over conventional methods that involve hazardous and corrosive reagents harmful to both the environment and human health. Spent LiBs treated with DESs need to be gently agitated, microwaved or sonicated to allow the DESs to penetrate the binder matrix. This process breaks the bonds between the binder matrix and the electrode materials, leading to the dissolution of the binder in the DES solution and its separation from the electrode materials.

Table 1 Recovery of valuable metals from the leaching of spent LiBs using various DES, along with their respective conditions and effectiveness

Condition	Leaching agent		Efficiency (%)						References
	HBA	HBD	Li	Ni	Co	Mn	Cu	Al	
T = 100 °C t = 60 min	Dimethyl-beta-propiothetin (DMPT)	Ethylene glycol (EG)	✓ 99.59	✓ 99.28	✓ 99.04	✓ 99.45			[8]
25.7 wt% water T = 120 °C t = 4 h	Choline chloride (ChCl)	Citric acid (CA)	✓ >97		✓ >97				[15]
T = 160 °C t = 360 min	Betaine hydrochloride (BeCl) and	Formic acid (FA)	✓ 98.03	✓ 92.35	✓ 94.19	✓ 96.01			[7]
T = 80 °C t = 3 min	Tetraethylammonium chloride (TEAC)	L-Ascorbic acid (LAA)	✓ 99.1%	✓ 99.6	✓ 99.4	✓ 99.3			[17]
T = 100 °C t = 80 min	Choline chloride (ChCl)	Phenylphosphinic acid (2PPA)	✓ 97.7%	✓ 97.0	✓ 96.4	✓ 93.0			[2]
T = 40 °C t = 6 h	Choline chloride (ChCl) With presence of O_3 oxidizing atmosphere	Ethylene glycol (EG)	✓ 91.1						[12]
T = 80 °C t = 5 h	Choline chloride (ChCl)	Oxalic acid (OA)	✓ 20.0	✓ 99.28	✓ 100.0	✓ 100.0			[14]
T = 180 °C t = 10 s (T = 110 °C t = 1 h)	Choline chloride (ChCl)	Oxalic acid (OA)	✓ 100.0		✓ 100.0				[6]
T = 140 °C t = 16 h	Choline chloride (ChCl)	Succinic acid (SA)/ethylene glycol (EG)			✓ 99.62				[4]
T = 105 °C t = 24 h	Choline Chloride (ChCl)	Lactic acid (LA)	✓ 100.0	✓ 100.0	✓ 100.0	✓ 100.0			[10]
T = 110 °C t = 360 min	Sulfosalicylic acid dihydrate (SAD)	Ethylene glycol (EG)	✓ 100.0	✓ 99.1	✓ 94.8	✓ 100.0			[13]

Previous studies reported that Al foil was successfully separated from cathode materials of spent LiBs with a yield of 99.86 wt% using a mixture of DES; choline chloride and glycerol (ChCl:Gly) at a molar ratio of 2.3:1 at 190 °C for 15 min [16]. The results indicated that the Al foil detached from cathode materials due to the deactivation of polyvinylidene fluoride (PVDF) through a high-temperature alkaline decomposition process within the DES.

Other study demonstrated that the potassium carbonate–ethylene glycol (K_2CO_3–EG) DES was effectively degrading the PVDF binder and separating the electrode active of spent LiBs [3]. The result showed that the reaction of the carbonate ions with the PVDF polymer chains led to the degradation of the polymer that leading to effective separation. This study reported that the electrode materials could be fully separated under mild reaction conditions (100 °C, 20 min), and the DES could be recovered and used again for several cycles. On the other hand, the use of choline chloride (ChCl):ethylene glycol (EG) of in 1:2 molar ratio, assisted by microwave treatment, successfully decomposed 100% of the binder in the cathode materials of spent LiBs [5]. This treatment was conducted at 120 °C, with duration of 8 min and the DES mixtures can be recycled up to nine times.

Even though the used DES can be recovered and reused several times, it is necessity to perform the dissolution condition at low temperature and within a short processing time, while maintaining effective degradation of the PVDF binder and separation of the electrode materials from spent LiBs. Additionally, density functional theory (DFT) calculation may be necessary to significantly enhance the understanding of the PVDF dissolution mechanism from THE electrode materials of spent LiBs.

5 Anode Graphite Recovery and Regeneration from Spent LiBs in DESs

Conventional treatment method to recover and regenerate the anode graphite from spent LiBs is associated with environmental pollution and ineffective impurity removal. Thus, the use of DES–choline chloride (ChCl) and glycerol (Gly) with molar ratio of 1:2 DESs was successfully eliminated impurities from spent graphite treatment at 175.5 °C as well as restoring its structure and morphology. DES possess strong hydrogen bond formation and high dissolution capacity which allows its penetration to the structure of the spent graphite. This penetration ultimately enhances the dissolution process and facilitates the removal of impurities, solid electrolyte interface (SEI) film and residual electrolytes. Furthermore, the lattice structure and morphology of recovered graphite can be restored as DESs be able to act as a medium of reordering the graphite's carbon atoms. Additionally, the DES effectively eliminate of the surface contaminants and defects through its chemical reaction or surface adsorption processes subsequently yielded cleaner and more active surface on the regenerated graphite. Thus, this regenerated graphite can straightly use in building a new battery anode.

6 Conclusion

As the demand for lithium-ion batteries continues to rise, so does the need for effective and sustainable recycling methods. The removal of heavy metals from spent batteries is crucial for minimizing environmental pollution, conserving natural resources, and fostering a circular economy. While mechanical, chemical, biological, and thermal methods each have their advantages, they also face significant challenges that must be addressed to make recycling more efficient and widely accessible.

Future efforts should focus on developing integrated recycling systems that combine the strengths of different methods while minimizing their drawbacks. Policy makers and industries must also work together to create incentives for recycling and invest in research to improve recovery technologies. By doing so, society can ensure the responsible management of lithium-ion batteries and support a more sustainable future.

Utilization of DESs in early stage of LiBs recycling might be costly compared to conventional acidic and organic solvents. However, this cost is outweight with its benefits such as reusability, less environmental impact and improve process efficiency in recycling spent LiBs. Even though the DES can be reutilize for several cycles, thorough studies are still needed to determine any decomposition of DES components, especially when used under harsh conditions (such as high temperature and presence of reactive materials). Such studies are crucial before implementation on an industry scale to minimize the risk of lower extraction yields and ensure recovery of valuable material. Furthermore, cost incurred and appropriate recovery and purification processes of used DESs are important too to ensure the DES properties are still intact for reuse over multiple cycles. Thus, it is of utmost important to evaluate DES's stability and composition to ensure its sustainability as alternative solvents for recycling LiBs, which sometimes need to operate under harsh conditions.

References

1. Abbott, A.P., Capper, G., Davies, D.L., Rasheed, R.K., Tambyrajah, V.: Novel solvent properties of choline chloride/urea mixtures†. Chem. Commun. **1**, 70–71 (2003)
2. He, X., Wen, Y., Wang, X., Cui, Y., Li, L., Ma, H.: Leaching NCM cathode materials of spent lithium-ion batteries with phosphate acid-based deep eutectic solvent. Waste Manag. **157**, 8–16 (2023). https://doi.org/10.1016/j.wasman.2022.11.044
3. Hua, Y., Xu, Z., Zhao, B., Zhang, Z.: Efficient separation of electrode active materials and current collector metal foils from spent lithium-ion batteries by a green deep eutectic solvent. Green Chem. **24**(20), 8131–8141 (2022). https://doi.org/10.1039/d2gc02844a

4. Huang, F., Li, T., Yan, X., Xiong, Y., Zhang, X., Lu, S., An, N., Huang, W., Guo, Q., Ge, X.: Ternary deep eutectic solvent (DES) with a regulated rate-determining step for efficient recycling of lithium cobalt oxide. ACS Omega **7**(13), 11452–11459 (2022). https://doi.org/10.1021/acsomega.2c00742
5. Li, P., Xu, H., Luo, S., Wang, Y., Zhang, L., Lin, Y., Li, Z., Guo, J., Xu, Y., Zhang, Y., Yue, Y.: Green and non-destructive separation of cathode materials from aluminum foil in spent lithium-ion batteries. Sep. Purif. Technol. **338**, 126625 (2024). https://doi.org/10.1016/j.seppur.2024.126625
6. Li, T., Xiong, Y., Yan, X., Hu, T., Jing, S., Wang, Z., Ge, X.: Closed-loop cobalt recycling from spent lithium-ion batteries based on a deep eutectic solvent (DES) with easy solvent recovery. J. Energy Chem. **72**(May), 532–538 (2022). https://doi.org/10.1016/j.jechem.2022.05.008
7. Liu, C., Yu, J., Hu, J., Xu, J., Yu, A., Liu, T., Wang, Z., Luo, X., Deng, C., Luo, F., He, J., Zeng, G.: Ultra-low viscosity betaine hydrochloride-formic acid deep eutectic solvent for leaching critical metals from spent NCM lithium-ion batteries. J. Environ. Chem. Eng. **12**(3), 112586 (2024). https://doi.org/10.1016/j.jece.2024.112586
8. Luo, Y., Deng, Y., Shi, H., Yang, H., Yin, C., Ou, L.: Green and efficient recycling method for spent Ni-Co-Mn lithium batteries utilizing multifunctional deep eutectic solvents. J. Environ. Manag. **351**, 119814 (2024). https://doi.org/10.1016/j.jenvman.2023.119814
9. Meng, J., Wang, Y., Zhou, Y., Chen, J., Wei, X., Ni, R., Liu, Z., Xu, F.: Development of different deep eutectic solvent aqueous biphasic systems for the separation of proteins. RSC Adv. **9**(25), 14116–14125 (2019). https://doi.org/10.1039/c9ra00519f
10. Morina, R., Callegari, D., Merli, D., Alberti, G., Mustarelli, P., Quartarone, E.: Cathode active material recycling from spent lithium batteries: a green (circular) approach based on deep eutectic solvents. ChemSusChem **15**(2) (2022). https://doi.org/10.1002/cssc.202102080
11. Smith, E.L., Abbott, A.P., Ryder, K.S.: Deep eutectic solvents (DESs) and their applications. Chem. Rev. **114**, 11060–11082 (2014)
12. Tang, S., Yang, Z., Zhang, M., Guo, M.: A simple green method for in-situ selective extraction of Li from spent $LiFePO_4$ batteries by synergistic effect of deep-eutectic solvent and ozone. Environ. Res. **239**, 117393 (2023). https://doi.org/10.1016/j.envres.2023.117393
13. Tang, S., Zhang, M., Guo, M.: A novel deep-eutectic solvent with strong coordination ability and low viscosity for efficient extraction of valuable metals from spent lithium-ion batteries. ACS Sustain. Chem. Eng. **10**(2), 975–985 (2022). https://doi.org/10.1021/acssuschemeng.1c06902
14. Thompson, D.L., Pateli, I.M., Lei, C., Jarvis, A., Abbott, A.P., Hartley, J.M.: Separation of nickel from cobalt and manganese in lithium ion batteries using deep eutectic solvents. Green Chem. **24**(12), 4877–4886 (2022). https://doi.org/10.1039/d2gc00606e
15. Wang, C., Ai, T., Gao, X., Lu, J., Liu, J., Zhu, W., Luo, Y.: Effective recycling of critical metals from $LiCoO_2$ batteries by hydrated deep eutectic solvents: performance, kinetic and mechanism. J. Water Process Eng. **59**, 105088 (2024). https://doi.org/10.1016/j.jwpe.2024.105088
16. Wang, M., Tan, Q., Liu, L., Li, J.: A low-toxicity and high-efficiency deep eutectic solvent for the separation of aluminum foil and cathode materials from spent lithium-ion batteries. J. Hazard. Mater. **380**, 120846 (2019). https://doi.org/10.1016/j.jhazmat.2019.120846
17. Zhang, T., Dao, J., Wang, J., Wan, R., Li, C., Wang, D., Bao, C., Zhang, Y., Zhang, Z., Dong, P.: Efficient leaching of valuable metals from NCM cathode materials by green deep eutectic solvent. J. Clean. Prod. **438**, 140636 (2024). https://doi.org/10.1016/j.jclepro.2024.140636

Deep Eutectic Solvents in Biotransformation Processes

Dzun Noraini Jimat and Sharifah Shahira Syed Putra

Abstract The rising demand for eco-friendly and sustainable technologies in product formulation and industrial processes has led to increased interest of biotransformation processes. Conventionally, organic solvents are used in biocatalysis reactions, however, they have several drawbacks such as high volatility, inflammable and inhibition of biocatalyst activity. Ionic liquids (ILs) have also been employed in biocatalysis and demonstrate good performance, but their environmental impact remains a concern. As an alternative, deep eutectic solvents (DESs)—mixtures of a quaternary ammonium salt with a metal salt or hydrogen bond donors (HBDs)—offer advantages such as low toxicity, non-flammability, low vapor pressure, biocompatibility, biodegradability, and tunability via the alteration of salt and HBD combinations. Numerous studies have shown that the use of DESs in biotransformation represents a promising approach for enhancing reaction efficiency, improving enzyme stability, and supporting sustainable industrial practices. Thus, this chapter discusses the roles of DESs in biocatalysis, how the physicochemical properties of DESs affect biotransformation, and a comparison with conventional solvents.

1 Introduction

The rising demand for eco-friendly and sustainable technologies in product formulation and industrial processes leads to the interest of biotransformation processes. Biotransformation can be defined as chemically or biologically transformation of substances which is mediated by the microbial enzymes [37]. It can also be known as a sequence of chemical reactions that occur in a compound (i.e. drug) because of enzymatic or metabolic activities by a microorganism [37]. Lastly, it can be defined

D. N. Jimat (✉)
Department of Chemical Engineering and Sustainability, Kulliyyah of Engineering, International Islamic University Malaysia (IIUM), Kuala Lumpur, Malaysia
e-mail: jnoraini@iium.edu.my

S. S. Syed Putra
Faculty of Industrial Sciences and Technology, Universiti Malaysia Pahang Al-Sultan Abdullah (UMPSA), 26300 Gambang, Pahang, Malaysia

D. N. Jimat (ed.), *Deep Eutectic Solvents*, Engineering Materials,
https://doi.org/10.1007/978-981-95-1912-5_4

as a chemical modification of a substance due to the enzymatic reaction within the body.

Conventionally, organic solvents are used in biocatalysis reactions, however, they have several drawbacks such as high volatility, inflammable and inhibition of biocatalyst activity [36]. Alternatively, ionic liquids (ILs) are used in biocatalysis and exhibit good performance but their environmental impact remains a concern [36]. On the other hand, deep eutectic solvents (DESs), mixtures of a quaternary ammonium salt with a metal salt or hydrogen bond donors (HBDs) are considered low-toxicity, non-flammability, low vapour pressure, biocompatibility, biodegradability and easy to manipulate via alteration of the mixtures of salts and HBDs. Thus, it is a good alternative to conventional solvents (such as hexane, acetone, methanol and etc.) in biotransformation processes. Many studies have reported that DESs are able to enhance biocatalysis via diverse mechanisms such as improvement of substrate solubility and enhance enzyme stereoselectivity [21, 36]. Besides, their ability to improve enzyme activity and stability in reaction media [36], enhance reaction yield and more sustain as the DESs can be recycled and reused [26].

Several biotransformations studies have been established in DESs to synthesize high-value chemicals using hydrolases, oxidoreductases, lyases, and amine transaminases, wherein DESs act as solvents, cosolvents, additives, or even as substrates [6, 27, 36, 38]. DESs need to be appropriately customized to ensure its compatible in biocatalytic applications such as amount of water to maintain enzyme activity, polarity to solubilize reactants and viscosity [38]. These deliberations can be done by introducing water, manipulating DESs compositions and its polar ratios as well as increasing temperature [38, 39]. In addition, immobilized biocatalysts are also employed to accommodate harsh conditions of DESs and it is much easier to process and re-used in DES systems [38].

Since DESs has low temperature freezing point (T_f) thus they keep in liquid form consequently appropriate to be used in biocatalysis reactions as monophasic or biphasic biocatalytic reactions occur in liquid environment. A deep eutectic system denotes to a eutectic mixture with a melting point much lower than either of the individual components. DESs mixture such as ChCl/urea with mol ratio of 1:2 can be used at room temperature as this liquid mixture has melting point of 12 °C [36]. At ambient temperature, a clear and homogeneous liquid is formed by a different group of DESs, recognized as natural deep eutectic solvents (NADES), which contain sugars, fructose, glucose, and amino acids with a molar ratio of 1:1:1 [36]. Furthermore, DESs are safe to be used compared to conventional solvents as their individual components are usually solids or inflammable liquid and exhibit very low volatility. Their simple, easy preparation and low toxicity also attracts its usage compared to ILs in biotransformation. Besides, DESs can be also considered biodegradable as most of the components are natural products such as NADES [36]. The biodegradability property of DESs depends on theirs HBAs and HBDs structure (i.e. number of –OH groups) and choline chloride/urea (ChCl/U) and choline chloride/acetamide (ChCl/A) are two examples of DESs which exhibit degradability about 80% [34].

2 Roles of Deep Eutectic Solvents (DESs) in Biocatalysis

Biocatalysis, which involves reactions catalyzed by biocatalysts such as isolated enzymes and whole cells, has seen remarkable advancements in both biocatalyst development and reaction media over the recent years. The structure of the enzyme is more stable when more hydrogen bonds are created when more water are present in DES. Furthermore, there are remarkable interactions exist between DES and the enzyme surface residues that leading to conformational stability of the enzyme. On the other hand, the contact between specific critical residues and the DES may cause the enhancement of enzyme activity. It is reported that the number of hydrogen bond donors increases the likelihood of interaction [17]. Khodaverdian et al. [17] studied the biochemical characterization, thermal stability and tertiary structure of laccase from Bacillus in betaine and choline-based NADES. They reported that the enzyme activity and stability were dramatically decreased in eutectics containing choline compared to betaine-based NADES. The highest enzyme activity was observed in 20% (v/v) glycerol:betaine (2:1). The highest enzyme stability was observed in sorbitol:betaine:water (1:1:1) and glycerol:betaine (2:1) compared to the aqueous buffer at 80 and 90 °C. They concluded that the variation NADESs used in this study posed different physic-chemical properties (viscosity, conductivity, polarity, surface tension and etc.) which may affect changes in Kcat. Furthermore, the presence of copper in the laccase enzyme could also give impact to the laccase behavior in NADESs.

Previous studies also found that the inclusion of NADES to the assay medium showed significantly increased the enzymatic kinetic parameters such as Michaelis constant (K_m) and rate constant (K_{cat}) values [17]. NADESs with HBDs such as sugars (sorbitol) or carboxylic acids (e.g., malic acid, citric acid) exhibit high viscosities which could be due to the presence of high intermolecular hydrogen-bond network [40]. A higher Km value is resulting from the binding affinity between substrate and the enzyme active site causes by the hydrogen bond interactions between the substrate and HBD component of natural deep eutectics [17]. At ambient temperature, NADES showed considerably high viscosities (>100 cP) [4].

Other study demonstrated improvement in lipoxygenase (from *Basidiomycete Pleurotus sapidus* (LOXPSA)) activity and stability in NADESs [11]. They found that betaine:sorbitol:water (1:1:3, BSorbW) and betaine:ethylene glycol (1:3, BEtGly) showed the greatest influence on the peroxidation of linoleic acid and the side reaction of piperine to the vanilla-like scented compound piperonal [11]. In comparison to the control, the yield of piperonal in NADESs enhance by 40% in BEtGly and 43% in BSorbW. Furthermore, the presence of BSorbW improved the enzyme's stability at different temperatures and raised its activity when incubated at 60 °C.

The viscosity of NADES is influenced by its HBD and HBA, which subsequently impact its interaction with proteins. Sugar-based NADESs exhibit strong polarity due to presence of more hydroxyl groups that give impact to the enzyme activity and stability. DES consists of primary metabolites, such as amino acids, organic acids, sugars, or choline derivatives, able to serve as a cytoplasm-like natural environment for enzymes promoting biotransformation.

It was found that combinations of betaine and ethylene glycol exhibited positive effects, specifically on the relative LOX-PSA protein activity. It also showed that co-solvents, such as NADESs or ionic liquids, may affect enzymatic stability and activity through modifications of the chemical structure, polarity, viscosity, and their ability to form hydrogen bonds with the protein surface. The protein-solvent bond is stable, and the enzyme has more hydrogen bonds with the solvent molecules and a more compact surface area that is accessible to the solvent.

Recently, deep eutectic solvents (DESs) have been utilized to enhance drug solubilization, specifically for drugs used in oral liquid formulations. Previous study reported that carvedilol, a drug that used for chronic hypertension was solubilized and stabilized in 11 eutectic systems constructed by choline chloride, water and an HBD (glucose, xylitol, sorbitol, or citric acid) [22]. The study observed that the solubility of carvedilol in all solvents was increased compared to that of aqueous solubility [22]. This result will facilitate the oral liquid drugs formulation for paediatric and geriatric peoples, who may encounter problem in swallowing.

Another study demonstrated that curcumin exhibited significant dissolution in NADES composed of glycerol and choline chloride [15]. This study also showed that the same solvents were effectively used as an extraction medium to obtain curcuminoids from original sources, as well as acting as an effective stabilizer in preventing the degradation of curcumin from sunlight [15].

The initial biotransformation study using deep eutectic solvents (DESs) as a solvent or co-solvent was carried out by Gorke et al. [12]. They studied the hydrolase-catalyzed transesterification and found that urea, as a strong hydrogen-bond donor (HBD), when paired with Ch as a hydrogen-bond acceptor (HBA), does not denature hydrolases and maintains significant enzyme activity. This result showed that the hydrogen-bond network in a DES able to reduce the chemical potential.

3 How Physicochemical Properties DESs Affects the Biotransformation

DESs can be created by mixing hydrogen-bond acceptor (HBA; such as choline chloride) and hydrogen-bond donor (HBD; such as polyalcohol, sugar alcohol or organic acid or amide). It is essential to determine the physicochemical properties of these solvents such as density, viscosity and pH value which could synergetic affect the biocatalysis reactions. The density of DESs is greater than water; in range of 1.1–1.4 g/cm^3. Commonly, viscosity of all eutectic solvents is very high (200–500 mm^2/s at 40 °C) which could affect the mass transfer and other processes such as pumping, mixing and fluid flow. These physicochemical properties are also influenced by the molar ratio of DESs mixtures and the operational temperature. A high temperature may cause to higher molecular activity and mobility consequently increases the molar volume of the DESs and reduce their density. Additional of water can reduce the viscosity and density of DESs subsequently improve the mass transfer but it could disrupt the halide-HBD supermolecular complex. Polarity of DESs is also crucial as it determines their solubility in biocatalytic reactions. Organic acid-based DESs are the most polar, while sugar and polyalcohol-based DES are the least polar. pH is also another factor that influences biocatalytic reactions. A low pH of eutectic mixtures can be enhanced with increment of their water content. While a high pH of eutectic mixtures can be decreased with increasing of their water content. Optimal molar ratio of the HBAs and HBDs is also crucial as it influences the H-bond net of biocatalytic reactions. Figure 1 shows the rules of thumb in designing an effective biotransformation specifically biocatalytic processes involving deep eutectic solvents (DESs) as suggested by Panić et al. [26].

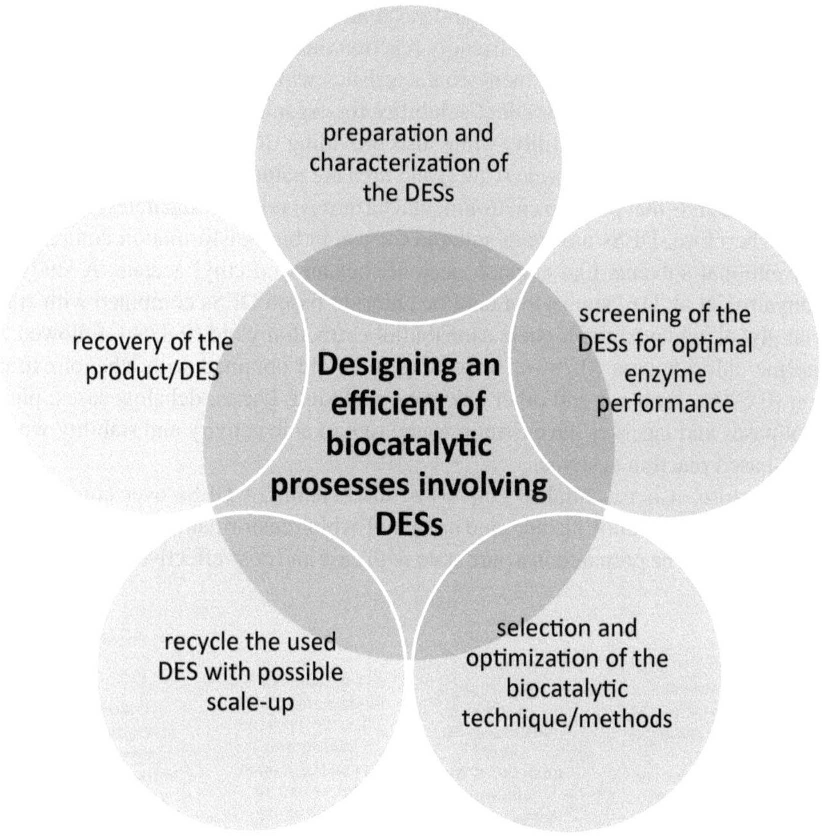

Fig. 1 Rules of thumb in designing an effective biotransformation specifically biocatalytic processes involving deep eutectic solvents (DESs) as suggested by Panić et al. [26]

4 Comparison of Biotransformation Between Conventional Solvent and DES

Deep eutectic solvents (DESs) can serve as solvents, cosolvents, or substrates in biocatalytic processes, utilizing whole cells or isolated enzymes to enhance selectivity and accelerate reaction rates [5]. Research has been conducted on the biotransformation using enzymes, cell extracts, and whole cells from yeasts, fungi, bacteria, microalgae, cyanobacteria, and plants. This biotransformation has involved both soluble and immobilized enzymes. In general, biotransformation refers to the structural modification of a compound through enzymatic chemical reactions occurring within a living system [37]. Additional definitions of biotransformation are presented in Fig. 2. However, the isolation and purification of the required enzymes can be challenging and costly [18]. Therefore, biotransformation using various types of DES can enhance the efficiency, selectivity, and sustainability of DES conversion processes by providing a tunable and eco-friendly reaction medium.

Additionally, DESs share many characteristics with conventional solvents, such as high thermal stability, excellent solubility for organic and inorganic substances, low toxicity, and non-volatility, while also providing distinct advantages [14]. For example, both choline chloride (ChCl) and urea are naturally occurring, biocompatible compounds that pose no environmental hazards if released separately into nature [13]. Therefore, DESs are more suitable for use in biotransformation compared to conventional solvents like ethanol, acetone, hexane and ethyl acetate. A study by Kanyairita et al. [16] shows that choline chloride-based DESs combined with ethylene glycol achieved the highest cannabidiol extraction yield (0.87%), followed by choline chloride:urea (0.78%), surpassing the yield obtained with ethanol extraction (0.69%). Protease and other enzymes, including lyases, dehalogenases, phospholipases and laccases have demonstrated remarkable activity and stability within DES-based reaction systems.

Therefore, it is essential to emphasize the benefits of DESs over conventional solvents such as ethanol, hexane, and methanol in biotransformation processes. Moreover, DESs can be prepared in a pure state with ease and cost-effectiveness, achieving

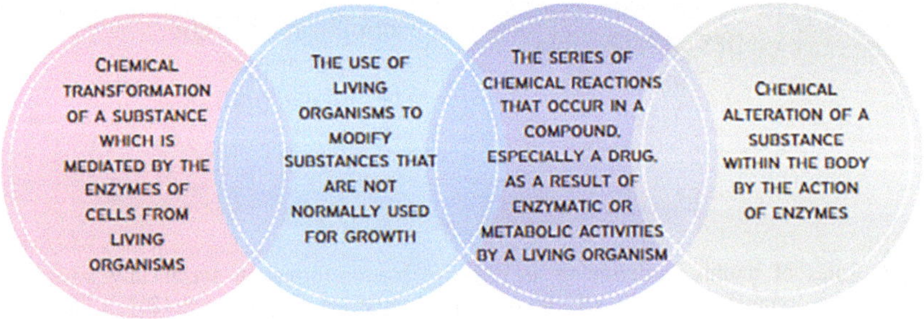

Fig. 2 Definitions of biotransformation to the structural modification of a compound through enzymatic chemical reactions

100% atom economy without requiring the post-synthesis purification typically needed for solvents such as ILs [10]. Several DESs, such as choline chloride:glycerol and choline chloride:ethanediol, have been documented for use in biocatalytic reactions [7]. For instance, a study by Khodaverdian et al. [17] examined laccase activity in DESs with varying hydrogen bond acceptor (HBA) and hydrogen bond donor (HBD) ratios. Their study showed that the chloride ion in ChCl inhibited laccase activity. A notable increase in laccase activity was observed when a 50% assay mixture of glycerol:betaine (threefold) and malic acid:betaine:water (1.5-fold) was used. In comparison, sorbitol:betaine:water (1:1:1) exhibited no significant changes relative to the buffer. Notably, ChCl–Gly DES demonstrated significant potential as a solvent for biotransformation reactions, achieving comparable or even superior conversion rates compared to toluene or hexane [3]. In general, DES can assist in reducing steric hindrance between the substrate and enzymes, while also addressing the high reduction potential of the substrate [1].

For instance, Toledo et al. [31] conducted molecular docking studies to investigate the interaction of laccase with various HBAs and HBDs at the molecular level. Their findings indicated that hydrogen bonds formed exclusively between all tested polyols and specific enzyme amino acids, particularly alanine, serine, and histidine. Furthermore, the docking affinity of laccase for HBDs showed a consistent increase with the order: ethylene glycol < glycerol < erythritol < xylitol. This trend was attributed to the rising number of hydroxyl groups in the polyols, which contributed to the enhanced laccase activity. A study by Ezgi Ünlü et al. [9] investigated the synthesis of polycatechin by laccase in DESs made from choline chloride:glycerol and betaine:mannose at concentrations of 5% and 50% (v/v). Their study demonstrated that DESs could successfully replace acetone in the polycatechin synthesis process. Despite their potential as designer solvents, the application of DESs as media for enzymatic reactions involving laccase requires further investigation. A more comprehensive understanding is needed to explore how DES compositions and their concentrations on aqueous solutions influence enzyme activity and stability.

Therefore, a study by Putra et al. [30] explores the potential of nanodiamonds (ND) as an innovative carrier for enzyme immobilization, evaluating their performance and comparing the efficiency of immobilized enzymes to their native counterparts. The findings reveal a marked enhancement in the efficiency of immobilized lipase, which preserved 76% of its initial activity after recycling with DES. This indicates that DES effectively stabilizes the enzyme's native structure during reuse. Toxicity assessments using brine shrimp demonstrated that ND immobilized with *Rhizopus niveus* lipase (RNL) showed negligible toxicity, even at low concentrations, suggesting its safety and biocompatibility. Additionally, Li et al. [19] successfully immobilized the enzymes 4LP-Tpebgl3@Na-Y and DtRha@ES-107 for the hydrolysis of total flavonoids of Epimedii (TFE) to produce icaritin using hydrophobic DES. Their study revealed that the relative enzyme activities of 4LP-Tpebgl3@Na-Y and DtRha@ES-107 were approximately 102.4% and 112.5%, respectively. Furthermore, the use of hydrophobic DES did not adversely affect the binding stability or thermal stability either enzyme.

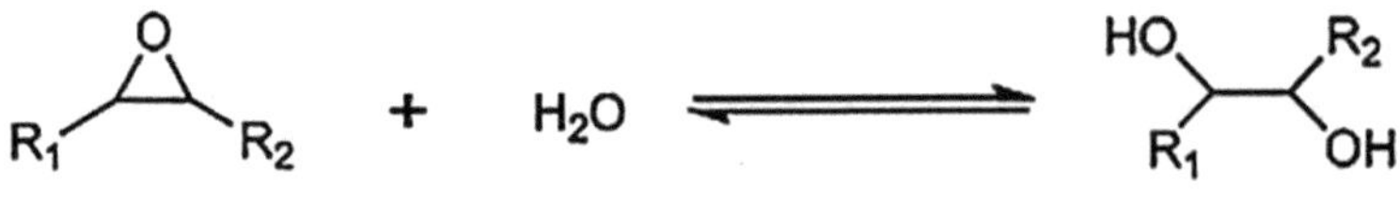

Fig. 3 **Scheme 1** Biocatalytic hydrolysis of epoxides

A study by Negi et al. [25] demonstrated a maximum conversion of 66.5% in the biotransformation of grease waste into fatty acids through submerged fermentation using *Penicillium chrysogenum* SNP5. This was achieved by optimizing media formulation and applying an artificial neural network (ANN) model, utilizing a conventional solvent mixture of n-hexane, diethyl ether, and acetic acid in a 50:4:1 ratio. However, a study by Mao et al. [23] successfully demonstrated the use of a DES composed of glycol and choline chloride in a 1:2 molar ratio, achieving an 82% enhancement in the system's performance for application in steroid hydroxylation using *Penicillium raistrickii*-mediated 15α-hydroxylation of D-ethylgonendione. It was observed that the use of DES can enhance the maximum conversion in *Penicillium*-based biotransformation processes by improving the solubility of substrates and providing a more favorable reaction medium through fermentation process. According to Lindberg et al. [20], incorporating DES as co-solvents (up to 25%) in aqueous solutions offers a promising platform for enhancing the activity of epoxide hydrolases. The next class of hydrolases employed in DES-compatible biotransformations is epoxide hydrolases. These enzymes play a crucial role in catalyzing the enantioselective hydrolysis of epoxides, leading to the formation of diols as shown in Fig. 3.

Besides that, Wang et al. [33] demonstrated that DES (ChCl:Urea 1:1) was the most effective co-solvent for enhancing the biotransformation efficiency of ω-transaminase (ω-TA). The specific activity of ω-TA in DES was enhanced by 3.9-fold due to strengthened interactions between the enzyme and the DES. Kinetic analysis indicated that DES significantly improved the enzyme's affinity for pyruvate and 1-(R)-PEA, increasing it by 1.63-fold and 1.93-fold, respectively, compared to the control in phosphate buffer. Another study using a different type of ω-TA demonstrated that the nickel binding sites on the surface of magnetic nanoparticles (MNP) enabled the direct immobilization of *Halomonas elongata* ω-transaminase (HeωT) in the presence of ethanol [18]. However, a loss of enzymatic activity was observed, which is a common issue in many immobilized enzyme systems. Thus, it shows that DES can enhance enzyme stability, activity, and selectivity in biotransformation processes, whereas conventional solvents, like ethanol, may lead to enzyme deactivation or reduced activity, especially in immobilized systems. For instance, a high ethanol concentration (30%) was necessary to maintain enzyme stability, with only a 12.7% reduction from its initial activity as reported by Pulido et al. [28]. However, its overall performance was not the most remarkable. This indicates that DES could be a more suitable and efficient alternative for specific biocatalytic applications.

A study by Chlipala et al. [2] identified an effective biocatalyst using *Yarrowia lipolytica*, which demonstrated significant activity in reducing C=C bonds through biotransformation with DES. Among the DESs tested, the lowest conversion was observed when glycerol served as the HBD. Additionally, a notable reduction in conversion was detected with an increase in the amount of added substrates. In contrast to the DES-based study with *Yarrowia lipolytica*, Tsirigka et al. [32] found that the methanol/hexane mixture can only achieved a maximum lipid extraction yield up to 27.6% per dry biomass (L/DB) and 12.1% L/DB using the conventional chloroform/methanol extraction method. However, a study by Drévillon et al. [8] demonstrated a successfully impact of high-pressure homogenization (HPH) on oil recovery from *Yarrowia lipolytica* yeast using *n*-hexane. They found that the maximum oil recovery using the wet method was about 80%. This was achieved by treating the cells, then extracting the oil by directly adding *n*-hexane and mixing with a high-speed disperser. Although the use of DESs offers advantages for biocatalysts like *Yarrowia lipolytica*, conventional solvents can also enhance efficiency, particularly in terms of extraction yield and reaction rate, depending on the specific application and process requirements.

In 2013, the application of ChCl-based DES for peptide synthesis under industrially viable conditions was demonstrated using α-chymotrypsin [24]. In this instance, adding 10–30% (v/v) water to the DES solution enhances enzymatic activity while minimizing product hydrolysis. Various ChCl-based DESs, utilizing xylitol, urea, isosorbide or glycerol as HBDs, were successfully employed, achieving high productivities of approximately 20 g L^{-1} day^{-1} at substrate loadings exceeding 100 g L^{-1} as demonstrated in Fig. 4.

Putra et al. [29] reported the successful immobilization of *Candida rugosa* lipase (CRL) onto nanocellulose (NC) extracted from almond shells by incorporating water into a DES system composed of choline chloride (ChCl), fructose, and water in a molar ratio of 5:2:5. They found that the DES with added water achieved a maximum specific activity of 6.57 U mg^{-1}, compared to 4.52 U mg^{-1} for the free CRL. Additionally, the immobilized CRL exhibited enhanced thermal stability, higher catalytic activity, and excellent reusability, maintaining its performance for up to seven consecutive cycles. Furthermore, Wu et al. [35] reported that DESs capable of providing the

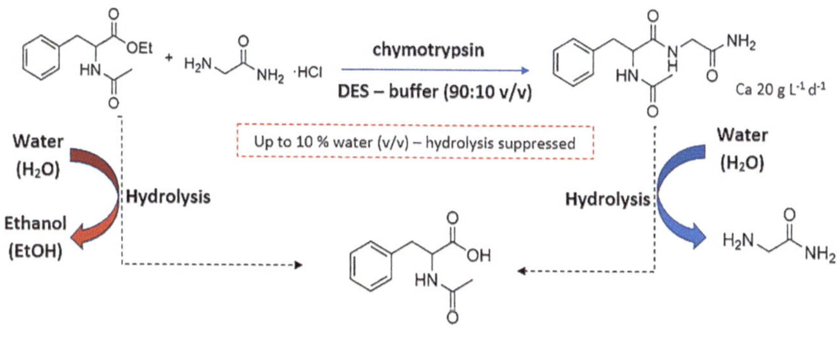

Fig. 4 Scheme 2 Chymotrypsin-catalyzed dipeptide synthesis in DES, achieving high productivity with suppressed hydrolysis at water content up to 20% (v/v) [24]

enzyme with a higher-helix content and a slightly more relaxed tertiary structure to facilitate the horseradish peroxidase (HRP) activity. This indicates that the presence of water can enhance the biotransformation efficiency of the enzyme, improving its activity, stability, and reusability.

Overall, the use of DES in biotransformation represents a promising approach for enhancing reaction efficiency, promoting enzyme stability, and supporting sustainable industrial processes. Table 1 demonstrated an overview of the use of DESs as media for biocatalytic reactions involving enzymes, cell extract and whole cells.

Table 1 An overview of the use of DESs as media for biocatalytic reactions involving enzymes, cell extract and whole cells

Enzymes/cell extract/whole cell	DES composition [molar ratio]	Results	References
Cannabidiol	Choline chloride:urea [1:2]	Highest cannabidiol extraction yield (0.87%)	Kanyairita et al. [16]
Laccase from *Bacillus* HR03 was	Glycerol:betaine [2:1], and malic acid:betaine:water [1:1:1]	Increase in laccase activity (malic acid:betaine:water (1.5-fold))	Khodaverdian et al. [17]
Laccase catalyzed polycatechin synthesis	Choline chloride:glycerol and betaine:mannose [50% v/v]	Successfully replace acetone in the polycatechin synthesis process	Ezgi Ünlü et al. [9]
Rhizopus niveus lipase (RNL)	Choline chloride:sucrose:water [4:1:4]	Preserved 76% of its initial activity after recycling with DES and negligible toxicity	Putra et al. [30]
Icaritin, a hydrolysate from total flavonoids of *Epimedii* (TFE)	Dodecyl alcohol:thymol [2:1]	The use of hydrophobic DES did not adversely affect the binding stability or thermal stability either enzyme	Li et al. [19]
Penicillium raistrickii-mediated 15α-hydroxylation of D-ethylgonendione	Glycol:choline chloride [1:2]	Achieving an 82% enhancement in the system's performance for application in steroid hydroxylation using *Penicillium raistrickii*-mediated 15α-hydroxylation of D-ethylgonendione	Mao et al. [23]
ω-Transaminase (ω-TA)	Choline chloride:urea [1:1]	The specific activity of ω-TA in DES was enhanced by 3.9-fold and DES significantly improved the enzyme's affinity for pyruvate and 1-(R)-PEA, increasing it by 1.63-fold and 1.93-fold, respectively	Wang et al. [33]

(continued)

Table 1 (continued)

Enzymes/cell extract/whole cell	DES composition [molar ratio]	Results	References
Culture of the yeast *Yarrowia lipolytica* KCh 71	Choline chloride:glycerol [1:2]	Demonstrated significant activity in reducing C=C bonds through biotransformation with DES especially when glycerol served as the HBD	Chlipala et al. [2]
Candida rugosa lipase (CRL)	Choline chloride (ChCl), fructose, and water [5:2:5]	DES with added water achieved a maximum specific activity of 6.57 U mg^{-1}, compared to 4.52 U mg^{-1} for the free CRL and maintaining its performance for up to seven consecutive cycles	Putra et al. [29]
Penicillium raistrickii	Glycol:choline chloride [1:2]	Successfully applied in steroid hydroxylation, and substrate bioconversion (72 h) enhanced to 82% in the system containing 4% DES, 2 g/L substrate and 17.5 gdw/L mycelium	Mao et al. [22]
Horseradish *peroxidase* (HRP)	HBA: Cholinium chloride (ChCl) and cholinium acetate (ChAc) HBD: Urea, glycerol, acetamide, ethylene glycol [1:2]	DESs capable of providing the enzyme with a higher-helix content and a slightly more relaxed tertiary structure may facilitate the horseradish peroxidase (HRP) activity	Wu et al. [35]

References

1. Chan, J.C., Paice, M., Zhang, X.: Enzymatic oxidation of lignin: challenges and barriers toward practical applications. ChemCatChem **12**(2), 401–425 (2020). https://doi.org/10.1002/cctc.201901480
2. Chlipała, P., Janeczko, T., Mazur, M.: Bioreduction of 4′-hydroxychalcone in deep eutectic solvents: optimization and efficacy with various yeast strains. Int. J. Mol. Sci. **25**(13) (2024). https://doi.org/10.3390/ijms25137152
3. Cvjetko Bubalo, M., Jurinjak Tušek, A., Vinković, M., Radošević, K., Gaurina Srček, V., Radojčić Redovniković, I.: Cholinium-based deep eutectic solvents and ionic liquids for lipase-catalyzed synthesis of butyl acetate. J. Mol. Catal. B Enzym. **122**, 188–198 (2015). https://doi.org/10.1016/j.molcatb.2015.09.005
4. Dai, Y., van Spronsen, J., Witkamp, G.J., Verpoorte, R., Choi, Y.H.: Natural deep eutectic solvents as new potential media for green technology. Anal. Chim. Acta **766**, 61–68 (2013). https://doi.org/10.1016/j.aca.2012.12.019
5. de Gonzalo, G., Martin, C., Fraaije, M.W.: Positive impact of natural deep eutectic solvents on the biocatalytic performance of 5-hydroxymethyl-furfural oxidase. Catalysts **10**(4) (2020). https://doi.org/10.3390/catal10040447

6. De María, P.D., Guajardo, N., Kara, S.: Enzyme catalysis: in DES, with DES, and in the presence of DES. In: Deep Eutectic Solvents: Synthesis, Properties, and Applications, pp. 257–271 (2019). https://doi.org/10.1002/9783527818488.ch13
7. Domínguez de María, P., Maugeri, Z.: Ionic liquids in biotransformations: from proof-of-concept to emerging deep-eutectic-solvents. Curr. Opin. Chem. Biol. **15**(2), 220–225 (2011). https://doi.org/10.1016/j.cbpa.2010.11.008
8. Drévillon, L., Koubaa, M., Vorobiev, E.: Lipid extraction from Yarrowia lipolytica biomass using high-pressure homogenization. Biomass Bioenerg. **115**(April), 143–150 (2018). https://doi.org/10.1016/j.biombioe.2018.04.014
9. Ezgi Ünlü, A., Prasad, B., Anavekar, K., Bubenheim, P., Liese, A.: Investigation of a green process for the polymerization of catechin. Prep. Biochem. Biotechnol. **47**(9), 918–924 (2017). https://doi.org/10.1080/10826068.2017.1365241
10. Ferreira, C., Sarraguça, M.: A comprehensive review on deep eutectic solvents and its use to extract bioactive compounds of pharmaceutical interest. Pharmaceuticals, **17**(1) (2024). https://doi.org/10.3390/ph17010124
11. Garbe, M., Lehmann, L.T., Berger, R.G., Ersoy, F.: Improvement in the stability and enzymatic activity of pleurotus sapidus lipoxygenase dissolved in natural deep eutectic solvents (NADESs). Life **14**(2) (2024). https://doi.org/10.3390/life14020271
12. Gorke, J.T., Srienc, F., Kazlauskas, R.J.: Hydrolase-catalyzed biotransformations in deep eutectic solvents. Chem. Commun. **10**, 1235–1237 (2008). https://doi.org/10.1039/b716317g
13. Halder, A.K., Cordeiro, M.N.D.S.: Probing the environmental toxicity of deep eutectic solvents and their components: an in silico modeling approach. ACS Sustain. Chem. Eng. **7**(12), 10649–10660 (2019). https://doi.org/10.1021/acssuschemeng.9b01306
14. Jagirani, M.S., Soylak, M.: Deep eutectic solvents-based adsorbents in environmental analysis. TrAC - Trends Anal. Chem. **157**, 116762 (2022). https://doi.org/10.1016/j.trac.2022.116762
15. Jeliński, T., Przybyłek, M., Cysewski, P.: Natural deep eutectic solvents as agents for improving solubility, stability and delivery of curcumin. Pharm. Res. **36**(8) (2019). https://doi.org/10.1007/s11095-019-2643-2
16. Kanyairita, G.G., Mortley, D.G., Boersma, M., Collier, W.E.: Comparison of the efficiency of deep eutectic and organic solvents in the extraction of phytochemicals from Cannabis sativa L. Separations **11**(4) (2024). https://doi.org/10.3390/separations11040106
17. Khodaverdian, S., Dabirmanesh, B., Heydari, A., Dashtban-moghadam, E., Khajeh, K., Ghazi, F.: Activity, stability and structure of laccase in betaine based natural deep eutectic solvents. Int. J. Biol. Macromol. **107**, 2574–2579 (2018). https://doi.org/10.1016/j.ijbiomac.2017.10.144
18. Lau, E.C.H.T., Dodds, K.C., McKenna, C., Cowan, R.M., Ganin, A.Y., Campopiano, D.J., Yiu, H.H.P.: Direct purifcation and immobilization of his-tagged enzymes using unmodifed nickel ferrite $NiFe_2O_4$ magnetic nanoparticles. Sci. Rep. **13**(1), 1–10 (2023). https://doi.org/10.1038/s41598-023-48795-x
19. Li, Q., Lu, S., Wu, X., Wang, L., Wang, Z., Zhao, L.: Application of hydrophobic eutectic solvent in efficient biotransformation of total flavonoids of Herba Epimedii. J. Biotechnol. **391**(June), 106–116 (2024). https://doi.org/10.1016/j.jbiotec.2024.06.007
20. Lindberg, D., de la Fuente Revenga, M., Widersten, M.: Deep eutectic solvents (DESs) are viable cosolvents for enzyme-catalyzed epoxide hydrolysis. J. Biotechnol. **147**(3–4), 169–171 (2010). https://doi.org/10.1016/j.jbiotec.2010.04.011
21. Logarušić, M., Šubar, K., Nikolić, M., Jurinjak Tušek, A., Damjanović, A., Radović, M., Radojčić Redovniković, I., Žnidaršič-Plazl, P., Kroutil, W., Cvjetko Bubalo, M.: Harnessing the potential of deep eutectic solvents in biocatalysis: design strategies using CO_2 to formate reduction as a case study. Front. Chem. **12**(October), 1–19 (2024). https://doi.org/10.3389/fchem.2024.1467810
22. Lomba, L., Polo, A., Martínez, N., Alejandre, J., Giner, B.: Solubility and stability of carvedilol in deep eutectic solvents: a step towards a sustainable pharmaceutical formulation in the liquid state. ChemistrySelect **9**(11) (2024). https://doi.org/10.1002/slct.202305218
23. Mao, S., Wang, X., Zhang, Z., Wang, S., Li, K., Lu, F., Qin, H.: 15α-Hydroxylation of D-ethylgonendione by Penicillium raistrickii in deep eutectic solvents DESs containing system. Biochem. Eng. J. **164**, 107781 (2020). https://doi.org/10.1016/j.bej.2020.107781

24. Maugeri, Z., Leitner, W., Domínguez De María, P.: Chymotrypsin-catalyzed peptide synthesis in deep eutectic solvents. Eur. J. Org. Chem. **20**, 4223–4228 (2013). https://doi.org/10.1002/ejoc.201300448
25. Negi, S., Anjum, F., Khare, S., Satyapriy: Biotransformation of grease waste into fatty acid by Penicillium chrysogenum SNP5 through media engineering and artificial neural network. Environ. Sci. Pollut. Res. **30**, 39653–39665 (2023). https://doi.org/10.1007/s11356-022-24990-7
26. Panić, M., Cvjetko Bubalo, M., Radojčić Redovniković, I.: Designing a biocatalytic process involving deep eutectic solvents. J. Chem. Technol. Biotechnol. **96**(1), 14–30 (2021). https://doi.org/10.1002/jctb.6545
27. Pätzold, M., Siebenhaller, S., Kara, S., Liese, A., Syldatk, C., Holtmann, D.: Deep eutectic solvents as efficient solvents in biocatalysis. Trends Biotechnol. **37**(9), 943–959 (2019). https://doi.org/10.1016/j.tibtech.2019.03.007
28. Pulido, I.Y., Prieto, E., Jimenez-Junca, C.: Ethanol as additive enhance the performance of immobilized lipase LipA from Pseudomonas aeruginosa on polypropylene support. Biotechnol. Rep. **31** (2021). https://doi.org/10.1016/j.btre.2021.e00659
29. Putra, S.S.S., Basirun, W.J., Elgharbawy, A.A.M., Hayyan, A., Hayyan, M., Mohammed, M.A.: Nanocellulose and natural deep eutectic solvent as potential biocatalyst system toward enzyme immobilization. Mol. Catal. **528**, 1–37 (2022). https://doi.org/10.1016/j.mcat.2022.112422
30. Putra, S.S.S., Chew, C.Y., Hayyan, A., Elgharbawy, A.A.M., Taskin-Tok, T., Hayyan, M., Ngoh, G.C., Saleh, J., Al Abdulmonem, W., Alghsham, R.S., Nor, M.R.M., Aldaihani, A.G.H., Basirun, W.J.: Nanodiamonds and natural deep eutectic solvents as potential carriers for lipase. Int. J. Biol. Macromol. **270**(P2), 132245 (2024). https://doi.org/10.1016/j.ijbiomac.2024.132245
31. Toledo, M.L., Pereira, M.M., Freire, M.G., Silva, J.P.A., Coutinho, J.A.P., Tavares, A.P.M.: Laccase activation in deep eutectic solvents. ACS Sustain. Chem. Eng. **7**(13), 11806–11814 (2019). https://doi.org/10.1021/acssuschemeng.9b02179
32. Tsirigka, A., Ntoula, M., Kontogiannopoulos, K.N., Karabelas, A.J., Patsios, S.I.: Optimization of solvent extraction of lipids from Yarrowia lipolytica towards industrial applications. Fermentation **9**(1), 1–18 (2023). https://doi.org/10.3390/fermentation9010035
33. Wang, H., Tao, Y., Masuku, M.V., Cao, J., Yang, J., Huang, K., Ge, Y., Yu, Y., Xiao, Z., Kuang, Y., Huang, J., Yang, S.: Effects of deep eutectic solvents on the biotransformation efficiency of ω-transaminase. J. Mol. Liq. **377** (2023). https://doi.org/10.1016/j.molliq.2023.121379
34. Wen, Q., Chen, J.X., Tang, Y.L., Wang, J., Yang, Z.: Assessing the toxicity and biodegradability of deep eutectic solvents. Chemosphere **132**, 63–69 (2015). https://doi.org/10.1016/j.chemosphere.2015.02.061
35. Wu, B.P., Wen, Q., Xu, H., Yang, Z.: Insights into the impact of deep eutectic solvents on horseradish peroxidase: activity, stability and structure. J. Mol. Catal. B Enzym. **101**, 101–107 (2014). https://doi.org/10.1016/j.molcatb.2014.01.001
36. Xu, P., Zheng, G.W., Zong, M.H., Li, N., Lou, W.Y.: Recent progress on deep eutectic solvents in biocatalysis. Bioresour. Bioprocess. **4**(1) (2017). https://doi.org/10.1186/s40643-017-0165-5
37. Yousuf, M., Mammadova, K., Baghirova, S., Rahimova, R.: Biotransformation: a one pot method of novel pharmacological importance mini review. Appro. Drug Des. Dev. **4**(4), 4–6 (2019). https://doi.org/10.19080/NAPDD.2019.04.555645
38. Zhang, N., Domínguez de María, P., Kara, S.: Biocatalysis for the synthesis of active pharmaceutical ingredients in deep eutectic solvents: state-of-the-art and prospects. Catalysts **14**(1) (2024). https://doi.org/10.3390/catal14010084
39. Zhang, N., Steininger, F., Meyer, L.E., Koren, K., Kara, S.: Can deep eutectic solvents sustain oxygen-dependent bioprocesses?—measurements of oxygen transfer rates. ACS Sustain. Chem. Eng. **9**(25), 8347–8353 (2021). https://doi.org/10.1021/acssuschemeng.1c03547
40. Zhang, Q., De Oliveira Vigier, K., Royer, S., Jérôme, F.: Deep eutectic solvents: syntheses, properties and applications. Chem. Soc. Rev. **41**(21), 7108–7146 (2012). https://doi.org/10.1039/c2cs35178a

Extraction of Active Compounds from Plants and Animals Using Deep Eutectic Solvents (DESs)

Amal A. M. Elgharbawy, Najihah Mohd. Noor, Huma Warsi Khan, Nor Azrini Nadiha Azmi, and Hamzah Mohd. Salleh

Abstract This section explores into the application of deep eutectic solvents (DESs) for isolating bioactive elements from both plants and animals, marking a fundamental transition towards greener and more efficient practices within the domain of green chemistry. Conventional extraction techniques frequently depend on solvents with significant environmental and health hazards due to their toxic and volatile nature. DESs stand out as a groundbreaking alternative, distinguished by their reduced melting points, harmlessness, and ability to decompose naturally, presenting a safer option compared to traditional solvents. The adaptability and effectiveness of DESs in isolating a wide range of bioactive substances are examined, as well as their potential to serve as a substitute for water in specific extraction scenarios. Additionally, the chapter addresses the obstacles tied to employing DESs, including viscosity and extraction specificity challenges, highlighting the necessity for continued research to refine and expand the use of DES-based methods for broader industrial use. Upcoming directions and breakthroughs in this area are considered, emphasizing

A. A. M. Elgharbawy (✉)
International Institute for Halal Research and Training (INHART), International Islamic University Malaysia, Kuala Lumpur, Malaysia
e-mail: amalgh@iium.edu.my

Bioenvironmental Engineering Research Centre (BERC), Department of Chemical Engineering and Sustainability, Kulliyyah of Engineering, International Islamic University Malaysia, Kuala Lumpur, Malaysia

N. Mohd. Noor
Department of Chemical Engineering and Sustainability, Kulliyyah of Engineering, International Islamic University Malaysia, Kuala Lumpur, Malaysia

H. W. Khan
Department of Chemical Engineering, Universiti Teknologi PETRONAS, Seri Iskandar, Perak, Malaysia

N. A. N. Azmi
Laboratory of Halal Science Research, Halal Products Research Institute, Universiti Putra Malaysia, Kuala Lumpur, Malaysia

H. Mohd. Salleh
Faculty of Halal Science and Sustainable Tourism and Halalan Thayyiban Research Centre, Sultan Sharif Ali Islamic University, Bandar Seri Begawan, Brunei Darussalam

D. N. Jimat (ed.), *Deep Eutectic Solvents*, Engineering Materials,
https://doi.org/10.1007/978-981-95-1912-5_5

tailoring DES compositions, merging them with cutting-edge extraction methods, and leveraging automation to boost procedural efficacy. The eco-friendliness of DESs is thoroughly evaluated via extensive life cycle analyses, stressing the need for their adoption to be in harmony with the fundamental objectives of green chemistry and the principles of a circular economy. This chapter situates DESs at the confluence of creativity, environmental responsibility, and operational effectiveness in the extraction of bioactive elements, indicating a fertile path for the evolution of green chemistry initiatives.

Keywords Deep eutectic solvents · Bioactive compounds · Green chemistry · Sustainable extraction · Life cycle assessment

1 Introduction

Water, known for its environmentally friendly attributes faces challenges in extraction processes due to inherent physical and chemical limitations. While it effectively extracts polar bioactive compounds, its efficiency lags behind organic solvents like methanol or ethanol, particularly for less polar, non-polar, and hydrophobic compounds [13]. The scientific community has turned its attention to ionic liquid (IL) as a new class of fluids, gaining recognition across various research fields since the early 2000s [33]. ILs, owing to their excellent properties, such as bonding abilities, tuneability, thermal stability, and biodegradability [34], find diverse applications in the separation, recovery and design of new bioactive compounds [35]. However, despite being classified as green solvents, ILs raise environmental concerns related to their synthesis, toxicity, and biodegradability [26].

To address these challenges, researchers have explored a new generation of solvents known as deep eutectic solvents (DESs), which have become a captivating subject of study, especially for extraction processes. DESs, characterized by eutectic properties arising from the combination of hydrogen bond donor (HBD) and hydrogen bond acceptor (HBA) [42], exhibit a lower melting point than their individual components, allowing them to exist in a liquid state at mild temperatures. This unique composition makes DESs versatile and applicable in extraction processes, offering a promising alternative. Furthermore, DESs have attracted significant attention as a category of solvents formed by combining two or more Lewis or Brønsted–Lowry acids and bases, known for their capacity to lower the freezing point compared to their separate constituents. The term "deep eutectic solvent" specifically refers to a eutectic mixture where the eutectic point temperature is lower than calculated from the Schröder–Van Laar equation applicable to an ideal liquid eutectic mixture [44]. This lower melting point, a result of the unique composition, is pivotal to the versatility and applicability of deep eutectic solvents, particularly in extraction processes.

In contrast to ILs, DESs have gained attention for their distinctive characteristics, providing a safer and environmentally friendly option [56]. DESs are known for their biodegradability and cost-effective synthesis in commercial quantities, providing a distinct advantage over ILs. The composition of DESs typically includes generally nontoxic, readily available, and sustainable compounds. Moreover, common DES constituents often consist of naturally occurring biocompatible compounds, posing no hazard when released into the environment [48]. Unlike traditional solvents used in extraction processes, which may raise environmental concerns due to toxicity and volatility, DESs offer a more sustainable option. This aligns with the increasing emphasis on green and sustainable practices within the facet of green chemistry, highlighting the potential of DESs to positively contribute to environmentally conscious extraction processes.

The use of environmentally friendly solvents, exemplified by DESs, becomes particularly crucial when considering the extraction of active compounds from plant and animal sources. These compounds, often referred to as bioactive compounds, play an important role in various applications across different fields. Traditional medicines, with roots reaching back through the ages, have consistently drawn upon the rich reservoir of natural products to effectively treat a myriad of diseases [31]. This historical reliance on nature's pharmacy seamlessly extends into modern times, where a substantial portion of pharmaceutical medications is derived from these natural sources, each housing a diverse array of bioactive compounds, each possessing unique therapeutic properties [18]. Over the past decades, there has also been a notable surge in the extraction and application of natural compounds of animal origins. Traditional Chinese medicines, embracing a spectrum of over 1,500 animal species, Unani medicine utilizing about 500 species of invertebrates, and Ayurvedic medicine relying on animal species and products, underscore the interdependence of fauna and flora in healing ailments [50]. These compounds, often used as dietary or therapeutic supplements, play an important role in preventing, alleviating, or treating diverse diseases and associated symptoms.

As we direct the landscape of extracting these bioactive compounds, the appeal of employing environmentally friendly solvents, particularly DESs, comes to the forefront. The conscientious use of DESs aligns with the broader context of sustainable and ethical practices. DESs, characterized by their eco-friendly nature, offer a promising alternative to conventional solvents. Their lower environmental impact, biodegradability, and cost-effective synthesis in commercial quantities provide a distinct advantage over other extraction methods, addressing concerns related to toxicity and environmental sustainability. This chapter comprehensively explores the extraction of active compounds from both plant and animal sources using DESs. The examination starts with a deep dive into the historical foundations, providing a contextual understanding of extraction techniques. Following this, the principles of extraction using DESs are elucidated, discovering into the unique aspects that make DESs suitable for this purpose. The chapter further explores applications in plant and animal-based extraction, emphasizing the versatility of DESs in different contexts. Sustainable aspects and environmental impact are analyzed, providing insights into how DESs contribute to eco-friendly extraction practices. Finally, future trends and innovations in the field are discussed, offering a forward-looking perspective on the continued evolution of extraction methodologies.

2 Historical Context of Extraction Techniques

The progression of extraction techniques in obtaining natural products from raw materials has witnessed significant advancement, combining traditional methods with modern technologies and a deeper understanding of scientific principles (Fig. 1). The primary objective of any extraction technique is to maximize the recovery of target compounds from a sample matrix while preserving the integrity of the molecules of interest and minimizing the co-extraction of other impurities or undesirable compounds [22]. At the core of this progress lies the extraction process, utilizing methods, such as solvent extraction, distillation, pressing, and sublimation [80]. Each method contributes to the diverse approaches in this field, with early extraction methods like maceration and solvent extraction holding historical significance for extracting bioactive compounds from plants and other natural materials [12]. Nonetheless, these traditional methods often involve time-consuming processes and substantial amounts of toxic organic solvents, posing safety concerns for workers and consumers [2].

The challenges associated with traditional extraction methods have led to a significant shift towards the adoption of modern and environmentally friendly extraction techniques. This transition represents a proactive response to address issues

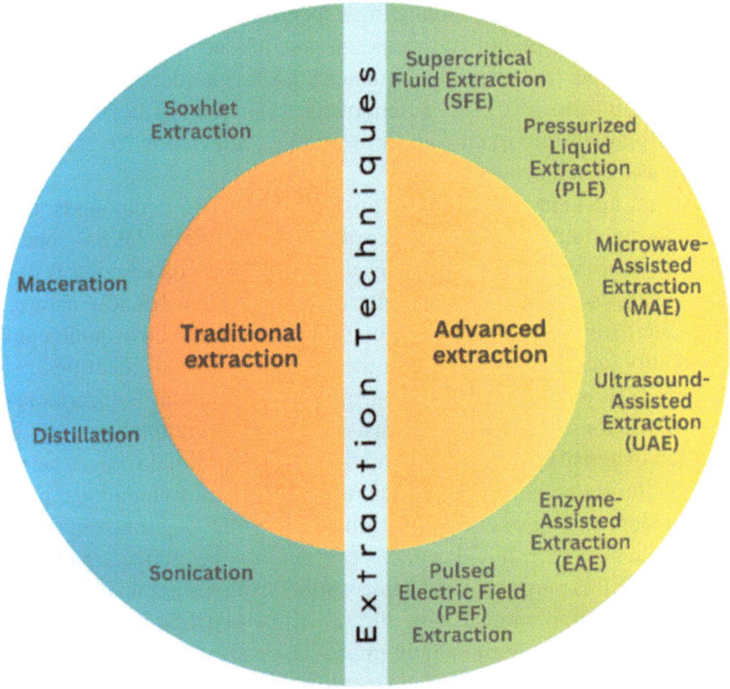

Fig. 1 Conventional and modern extraction techniques

related to safety concerns, environmental impact, and overall efficiency in the extraction process. Some modern or greener extraction methods, including supercritical fluid extraction (SFE), pressurized liquid extraction (PLE), and microwave-assisted extraction (MAE), have gained recognition in response to the increasing demand for sustainable extraction practices [80]. These advanced techniques signify a departure from conventional methods, offering numerous advantages aligned with the principles of green chemistry and environmental sustainability.

The choice of a suitable solvent for extraction plays an important aspect in the efficiency and safety of the process. Solvents play a critical role in dissolving target compounds from the sample matrix, and the selection process is influential in determining the overall outcome of the extraction [34, 58]. Consideration of factors, such as polarity, viscosity, and selectivity are essential during solvent selection, as these properties directly influence the extraction process [10, 74]. Due to their effectiveness, traditional solvents, including hydrocarbons and chlorinated solvents, have been widely used in extraction processes. However, their use raises notable environmental and safety concerns. However, their use raises notable environmental and safety concerns. Various solvents, such as hydrocarbons, aliphatic, esters, alcohols, halogenated solvents, and ketones, are commonly used in extraction processes. These solvents often originate from olefins, primarily ethylene and propylene, or crude oil. The appeal of these traditional solvents lies in their well-established nature, ready availability, and affordability compared to other raw materials used in various formulations. Despite their advantages, traditional solvents, originating mostly from olefins or crude oil, present environmental and safety challenges, prompting a growing interest in exploring alternative solvents [4]. While traditional solvents are renowned for their high solvency powers, rapid evaporation rates, and their ability to meet quality standards for the final product, their use is accompanied by concerns related to environmental impact and safety. Their potential harm to ecosystems and human health, coupled with issues related to their volatile nature, has led to an increasing interest in investigating alternative solvents [52]. This shift in focus has driven exploration into greener alternatives, such as DESs, which offer the potential to address the limitations associated with traditional solvents, particularly in terms of sustainability and environmental impact.

The historical development of DESs as extraction solvents can be traced back to the early twenty-first century, representing a relatively recent but rapidly advancing field in solvent technology. The conceptual foundation of DESs is rooted in the idea of eutectic mixtures, where the combination of two or more components yields a mixture with a lower melting point than the individual components. The eutectic concept was initially explored in the early twenty-first century, coinciding with the emergence of ILs. Historically, the concept of DES as a green alternative to ILs was introduced by Abbott and his co-workers with the production of the first generation of DESs. These early DESs were created using quaternary ammonium salts and amines or carboxylic acids as the hydrogen bond donors [1]. DESs were considered as a new generation of ILs, sharing similar physicochemical characteristics but with distinct differences [67]. The difference lies in their composition and the mechanism of their formation [56]. While ILs typically consist of an organic cation and an inorganic anion with relatively low melting temperatures, DESs are formed by combining two or more components, often involving a hydrogen bond donor and acceptor. This combination results in a eutectic mixture with unique properties, distinguishing DESs from ILs in terms of their composition and formation mechanism. In recent developments, primary metabolites have been combined with bio-renewable starting materials, such as sugar alcohols, simple sugars, amino acids, and organic acids, to create a subclass of DESs referred to as natural deep eutectic solvents (NADESs) [5]. Both DESs and NADESs have been considered the greenest nonaqueous solvents, depending on the components chosen for their preparation [78]. Hence, they are used in extraction processes (Fig. 2).

DESs are known for their remarkable ability to dissolve a broad spectrum of substances, ranging from polar to non-polar. This unique property positions DESs as efficient and environmentally friendly solvents, particularly in the extraction of bioactive compounds from medicinal plants [77]. Various classes of bioactive

Fig. 2 Evolution of deep eutectic solvents (DESs) in solvent extraction

compounds, such as flavonoids, phenolic acids, polyphenols, tanshinones, rosiglitazones, saponins, and anthraquinones, have been successfully extracted using DESs [79]. One of the early studies on the extraction of bioactive compounds dates back to 2015 when Nam et al. [51] focused on using a custom-made DES called PG-1 for efficiently extracting flavonoids from Flos Sophorae (*Sophora japonica*), a traditional Chinese medicine using tailor-made DESs. DESs are considered eco-friendly solvents, and PG-1, made from L-proline and glycerol, proved more effective than methanol in the extraction process [51]. The study optimized the extraction method using PG-1 and ultrasound assistance, achieving greener and more efficient results compared to conventional methods. The tailor-made extractant, PG-1, showed antioxidant activity due to its component, L-proline. In another study by Koraqi et al. [39], ultrasound assistance was integrated with a DES tailored based on citric acid for the extraction of bioactive compounds from Rosehip (*Rosa canina* L.) fruits. This study specifically aimed to explore the antioxidant activity of the extracted compounds, contributing valuable insights to the field. These DES included an HBD, such as glycerol and ethylene glycol, and an HBA like citric acid. The most efficient extract was obtained by utilizing a citric acid/ethylene glycol mixture through ultrasonic-assisted extraction. The study demonstrated a strong correlation between phenolics and anthocyanin contents in *Rosa canina* L. fruits, emphasizing the contribution of anthocyanins to overall phenolic content. In the latest study by Dardavilla et al. [14], the focus was on extracting compounds from *C. vulgaris* using DES-based solvents. Three choline chloride (ChCl) based DESs were tested, and the most efficient one, ChCl:1,2-butanediol (1:4)/H_2O 70/30 w/w, was identified. This solvent demonstrated optimal extraction of total carotenoid (TCC) and phenolic contents (TPC), along with strong antioxidant activity (IC_{50}). For further optimization, the extraction process using the identified solvent was studied by varying three parameters: solvent-to-biomass ratio, temperature, and time. The optimization process revealed that higher temperature and solvent-to-biomass ratio improved the extraction of carotenoids and phenolics, enhancing antioxidant capacity. The optimal conditions were found at 60 °C, a 40:1 solvent-to-biomass ratio, and an extraction time of about six and a half hours.

Apart from the extraction of bioactive compounds from plants, DESs and NADESs are also used in the extraction of marine sources. Astaxanthin (AXT) from crab shell wastes, shrimp shells, mussels, and *Haematococcus pluvialis*, were extracted using terpene based-NADESs [76]. The results indicate that methanol:myristic acid (8:1) NADES provided up to 60% higher AXT yields from crab shell residues compared to other NADESs. Moreover, using the same system at 60 °C for 2 h resulted in AXT yields comparable to a 6 h Soxhlet extraction with acetone. When applied to other biomasses, AXT yields increased significantly, up to 657-fold, compared to Soxhlet extraction. Another study investigated the use of betaine/polyol-based deep eutectic systems (DES) for extracting bioactive protein derivatives from sardine processing waste [59]. The goal was to maximize the extraction of target compounds, evaluating the impact of temperature, extraction time, and solid/liquid ratio. DES extraction results were compared to conventional water-based extraction. While DES and water showed similar protein extraction yields, DES extracts at 80 °C demonstrated significantly higher antioxidant and antimicrobial potential than water extracts. The increase in activity was up to threefold for antioxidants and over 250-fold for antimicrobials in the most promising DES extracts. The study suggests that DES not only serve as an efficient alternative for extracting bioactive compounds from fish biomass but also enhances the bioactivity of the obtained extracts. These findings collectively highlight the efficiency of DESs in the extraction process, offering a sustainable and effective alternative for obtaining bioactive compounds from various natural sources.

3 Principles of Extraction Using DESs

3.1 The Fundamental Principles of DES-Based Extraction

Deep eutectic solvents (DES) are distinguished by their ability to serve as extracting solvents for a wide variety of chemical solutes, such as flavonoids, pharmaceuticals, drugs, metals, etc. This is one of the attributes that set them apart from other materials. Viscosity, density, miscibility, polarity, bonding ability, tunability, solubilizing power, etc. are some of the essential physical properties that govern how effective they are in extraction methods. When it comes to enabling good mixing, it is advisable to choose solvents that have a low viscosity. While this is going on, an attempt is being made to achieve a significant density difference from the matrix to facilitate the following separation of phases [44]. Figure 3 shows the synthesis and properties of DES that make them suitable for extraction. These properties can be explained as follows:

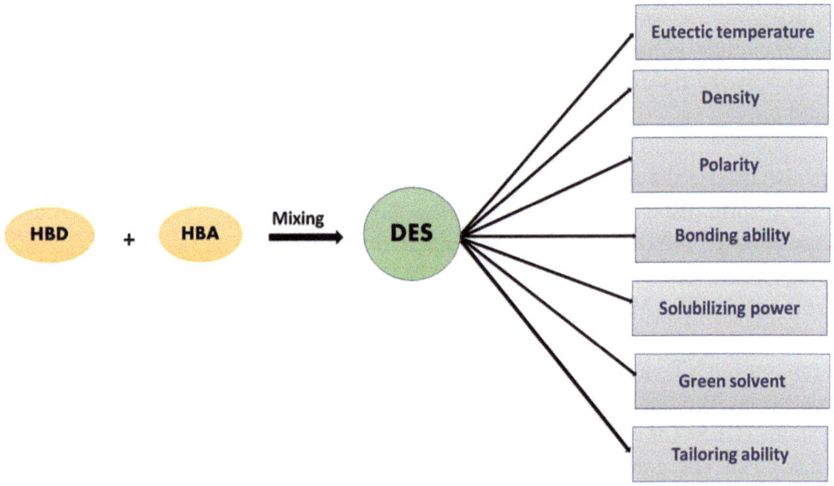

Fig. 3 Synthesis and properties of DESs

i. **Eutectic Point**: An outstanding point of DES is eutectic temperature. The lowest temperature at which the solvent's component mixture forms a eutectic mixture and has a distinct melting point is known as the eutectic temperature of a deep eutectic solvent (DES). A particular combination of materials with a lower melting point than any of its constituent parts is known as a eutectic mixture. The eutectic temperature of a deep eutectic solvent depends on the specific combination of components used to form it. Given that it influences both the chemical and physical characteristics of the deep eutectic solvent, the eutectic temperature is an important parameter. Depending on the components chosen, it might vary greatly and is determined by experimentation. To customize the characteristics of DES for particular uses, such as in green chemistry, as solvents for chemical reactions, as electrolytes in batteries, for active pharmaceutical ingredient (API) formulation, drug delivery, and pharma waste removal, researchers frequently experiment with diverse mixtures. For example, the [ChCl] and urea-based DES has a eutectic temperature of 304.95 K, whereas the actual melting point is 575 K and 406 K for [ChCl] and urea respectively [24].

ii. **Viscosity**: The resistance of a fluid to flow is measured by its viscosity. To ensure any separation process viscosity plays an important role. As DES is composed of HBD and HBA, viscosity depends upon the composition. The molar ratio of the components, the presence of additives, and the characteristics of the hydrogen bond donor (HBD) and hydrogen bond acceptor (HBA) can all have an impact on viscosity. The higher viscosity of DES is a bit challenging to be used for commercial applications. The model that is most frequently used to describe the viscous behaviour of DESs is the Arrhenius equation (Eq. (1)), which is expressed.

$$\acute{\eta}_{Arr} = Ae^{E/RT} \tag{1}$$

where E [kJ/mol]

R = molar gas constant

A = a prefactor, and the activation energy for viscous flow

T = temperature.

To overcome viscosity problems some researchers came up with the idea of adding water to DES. The maximum limit of adding water cannot be greater than 40%. It has been proposed that water content in DES can be used for tuning and controlling the properties [44].

iii. **Density**: Density is a significant parameter that may give information about the intermolecular interactions that occur in a DES. Overall, DESs have been shown to have densities that are higher than that of water. For instance, ethaline has a density of 1.14 g/cm^3, and glyceline has a density of 1.19 g/cm^3. They examined the density of ethylene, diethylene, triethylene glycol (TEG), and glycerol as HBDs, and they analyzed the density of benzylammonium chloride [Bam][Cl] salts as HBAs. Basaiahgari et al. recorded their findings. In comparison to DES based on glycerol, the density of DES based on ethylene glycol was found to be lower, as demonstrated by their findings. Increasing the amount of –OH functional groups on the HBD leads to the production of additional H-bonds, which in turn leads to a decrease in the free volume that is available. This is seen by the rise in density that occurred when ethylene glycol was replaced with diethylene, TEG, and glycerol [6]. The literature reveals that an increase in OH groups in HBD results in the formation of more hydrogen bonds thereby decreasing free volume. Also, sometimes the density of DES depends upon the amount or molar ratio of HBD and HBA. For example, the increase in the amount of choline chloride [ChCl] in a ChCl and citric acid [CA]-based DES decreases the density. Also, the density of DES depends on the temperature. With an increase in temperature, the density decreases.

iv. **Bonding ability**: An excellent property of DES is the hydrogen bonding ability of DES. This is related to the intermolecular force behind the melting point depression that takes place during DES synthesis, together with steric effects and ionic bonds between cations and anions pair [30].

v. **Solubilizing power**: DES has been used efficiently as a dissolution solvent cause of its high solubilizing power. Excellent solubilizing power is caused by their strong hydrogen bond interactions and thermal and chemical stability. One of DESs' biggest benefits is its tunability because of the physicochemical characteristics of extraction solvents—particularly polarity that have a significant impact on the extractive power. DES polarity can be altered based on the intended compound's characteristics [19]. DES made from [ChCl]:[OA] was successful at solubilizing highly insoluble marine samples [29].

vi. **Tunability**: DES is sometimes also referred to as tailored solvents. The reason underlying this is that DES can be designed and synthesized for specific solutes and specific applications. By the appropriate selection of HBD and HBAs the physico-chemical characteristics of DES can be altered. Polarity, hydrogen, bonding, density, viscosity, etc. can be altered easily [19]. The tunability and tailoring ability of DES makes them versatile for the extraction of active compounds from various sources [20].

3.2 Efficiency and Sustainability in DES Extraction Compared to Conventional Method

DES is usually composed of non-toxic components that are HBD and HBA. As discussed earlier, these components include quaternary ammonium, tetraalkylammonium, phosphonium, choline chloride, carboxylic acids, alcohols, etc. These DES starting material components overcome the challenge of toxicity of solvents during extraction processes. Furthermore, they are biodegradable and easily can be recycled. These solvent components are in line with the UN sustainable development goals [37], for the development of newer greener innovative processes and sustainable goal practices [36]. DES finds versatile applications in various extraction processes, such as liquid–liquid extraction (LLE), liquid-phase microextraction (LPME), microwave-assisted extraction (MAE) and solid-phase extraction (SPE). DES can be used for the extraction of bioactive compounds, metabolites, drugs, gases, and others. Table 1 presents some of the research using DES for the separation of different solutes using different extraction methods. Tang and his co-workers successfully utilized ChCl: EG-based DES for the extraction of terpenoids from cypress (*Chamaecyparis obtusa*) leaves [64]. Wang and co-workers explored the use of trioctylmethylammonium chloride [TMOAmCl] and decanoic acid [DA]-based DES for the extraction of analytes with 90–100% recovery efficiencies for the analytes understudy [72]. Khezeli and Daneshar utilized magnetic DES for the separation of thiophene. DES was composed of [ChCl], phenol [Ph] and $FeCl_4$ and [ChCl]:[EG] and $FeCl_4$. The results reveal that [Ph]-based DES was highly efficient and resulted in higher extraction efficiency of thiophene [38].

Table 1 DES-based extraction for various bioactive compounds

Solute	Extraction method	DES	Finding	References
Terpenoids	LPME	ChCl:EG	The small quantity of DES was efficient for the recovery of terpenoids	[64]
Analytes	UAE	[TOMAmCl]:[DA]	DES-based extraction was efficient cause the extraction and purification in a single step	[72]
Thiophene	UAE	[ChCl]:[Ph]:[An·$FeCl_3$]	Almost 100% separation was achieved. Also, the centrifugation step was not required and the DES can be recycled and reused	[38]
Active ingredients	MAE	[ChCl]:[Propane diol]	As compared to traditional extraction methods, MAE was efficient	[11]
Phenol	LLE	[TEAmCl]:[Ph]	DES resulted in a complete recovery of phenol	[28]
Quercetin	Extraction	[CA]:[Glu]:[H_2O]	DES enhanced the efficiency of quercetin from onions using the green extraction method	[25]

Twenty-five [ChCl]-based DES were efficient for the extraction of hydrophilic and hydrophobic components from radix in ultrasound-assisted dispersive liquid–liquid microextraction. The presence of DES improved and enhanced the extraction process [11]. Guo and his co-workers investigated the use of quaternary ammonium-based DES for the separation of [Ph] from model oil. The authors reported that DES composed of quaternary ammonium salts with appropriate alkyl chain lengths and anions with higher electronegativity were suitable for [Ph]. Almost 99.9% separation efficiency was achieved [28]. Gomez and his research team proposed NADES as greener solvents for the separation of quercetin. DES-based extraction was fast, sustainable, and eco-friendly for the separation of food-grade compounds [25]. Other studies of the DES-based extraction were listed in Table 1.

4 Applications in Plant-Based Extraction

Recently, there has been a notable shift towards adopting sustainable and environmentally conscious approaches in extracting bioactive compounds from plants, initiating a significant transformation in green chemistry. DES, an emerging category of environmentally friendly solvents, have attracted substantial interest due to their versatility, especially in the context of plant-based extraction. Positioned as a sustainable substitute for conventional solvents, DES is recognized for its enhanced efficiency, safety features, and reduced environmental impact. DESs emerge as potent tools for extractions and separations owing to their remarkable solvating capability for polar and non-polar substances. Notably, flavonoids, phenolic acids, and carotenoids which are crucial bioactive components, have been successfully extracted from various fruits and vegetables through the application of DESs [7].

Choline chloride-based natural deep eutectic solvents (NADESs) have been used for efficient extraction of total phenolic compounds (TPC) from leaves, pods and roots of an ornamental garden shrub, *Rhamnus alaternus* [53]. The leaves, pods and roots extracts obtained using the choline chloride–glycerol (ChCl–Gly) mixture exhibited a high value of TPC and possessed the best antioxidant properties with ABTS radical scavenging capacity, DPPH assay, reducing activity and iron chelation activity. The study suggested that the DES-based method developed was selective, efficient and sustainable for the extraction of polyphenols from plants. Additionally, the extraction method of rutin and rosmarinic acid based on DESs from *Satureja montana* L. has been reported [32]. The results showed that the quantities of extracted rutin and antiradical activity are more pronounced in extracts obtained through the application of DESs compared to organic solvents. Conversely, the content of rosmarinic acid in these extracts is less than that found in extracts obtained with organic solvents. Despite that, the study concludes that DESs present a more environmentally friendly option compared to their organic counterparts, and owing to their design flexibility and adjustable characteristics, they can be further tailored for the extraction of specific components.

In another study, a series of benzylammonium chloride-based-deep eutectic systems for the extraction of bioactive compounds from *Gardenia jasminoides* Ellis [75]. An optimal solvent which is dodecyldimethylbenzylammonium chloride–levulinic acid (1:3, mol/mol) with 35% (v/v) water was determined by response surface methodology experiments. In contrast to conventional methods employed in the past, a rapid, high-yield, and environmentally friendly extraction procedure was accomplished using straightforward heating conditions that did not impose limitations on the instrument. The study proved that DESs of dodecyldimethylbenzylammonium chloride–levulinic acid could serve as a sustainable and reusable solvent for the efficient extraction of natural bioactive compounds from plant-based raw materials.

A mixture of choline chloride (ChCl) and propylene glycol (PG) at the molar ratio 1:2 with 20 wt% water content was found optimal to extract RG-I type pectin from goji berry (*Lycium barbarum*) [27]. Scanning electron microscopy (SEM), X-ray diffraction (XRD), Raman spectra and density functional theory (DFT) calculations proved that the role of DES was to penetrate cell walls and interact with target substances. The effective extraction mechanism of the DES-water system was based on cell wall cellulose disassembly, side chain exposure, and high-branched pectin dissolution stages as shown in Fig. 4.

Extraction techniques are divided into conventional and innovative methods. This also applied to the utilization of DESs. Although conventional methods are simple and cost-effective, they have a long operational time, high energy consumption, and limited recovery of bioactive compounds due to the decomposition of thermally sensitive compounds. Besides the conventional method of maceration, the amalgamation of DESs with either microwave-assisted extraction (MAE) or ultrasound-assisted extraction (UAE) also brings excellent innovation in the extraction of plant-based compounds. This combination embraces the unique properties of DES, known for their efficiency in solvating a wide range of compounds, with the rapid and selective heating capabilities of the microwave or the cavitation effects induced by ultrasound. The result is a powerful and efficient extraction process that not only enhances

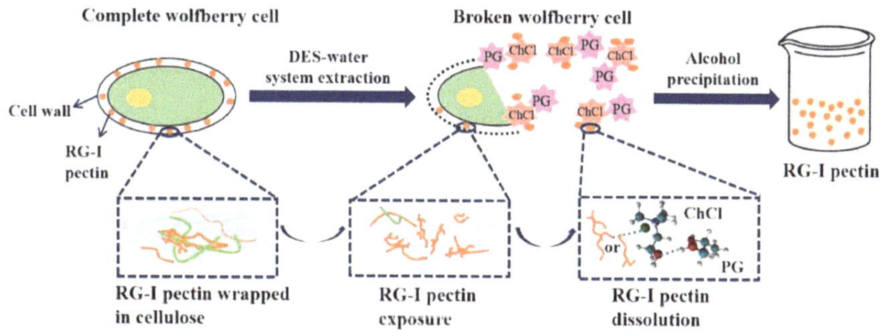

Fig. 4 The extraction mechanism of the DES-water system. Adapted from [27]

the yield of bioactive compounds but also aligns with the principles of green and sustainable extraction methodologies.

In comparison to traditional extraction procedures using organic solvents, this advanced methodology provides several advantages, including greater extraction efficacy, a lesser time required, and a low solvent demand. The extraction of pectin from dragon fruit using NADES with the microwave-ultrasound-assisted extraction method has been reported [66]. Among all possible methods and solvents, a combination of MUAE and NADES on pectin extraction from dragon fruit peels can favour a higher yield with desirable quality. NADES in combination with UAE and MAE, presented a higher extraction yield of phenolics and flavonoids from *Abelmoschus sagittifolius* (Kurz.) Merr. roots than that using a single method [69]. The advantages of NADES-based UAE might result from the material surface destruction of ultrasound and NADESs and the mass transfer improvement of microwave-based heating. These effects significantly improve extraction yield via mass transfer, reduce solvent consumption, and shorten extraction time. Doldolova et al. [16] prepared turmeric extracts by MAE using five NADES containing binary combinations of choline chloride, lactic acid, fructose, and sucrose. All NADES extracts, except the one containing fructose and choline chloride, exhibited higher TAC and CC than those in 80% methanol: water which was the preferred solvent in literature. NADES solvents did not interfere with subsequent antioxidant capacity measurements. Thus, it is evident in this study, that the proposed MAE and NADES are potentially efficient and sustainable in the pharmaceutical and food industries for the extraction of antioxidants and curcumin from turmeric.

Positioned as a sustainable alternative to conventional solvents, DES enhances efficacy, ensures safety, and minimizes environmental impact. The results from DES-based extractions in plants surpass those obtained with organic solvents, particularly in terms of biological activity, such as antioxidant potential, and other relevant parameters. The integration of DES reflects a transformative approach, aligning with green chemistry principles. Its potential to significantly contribute to mitigating environmental impact in extraction processes, coupled with improved efficiency and safety, positions DES as a promising solution in the pursuit of eco-friendly and effective plant-based compound extraction.

5 Applications in Animal-Based Extraction

Deep eutectic solvents are not limited to extracting substances from plants; their potential extends to the extraction process from animal sources. As a new class of environmentally friendly solvents, DES possesses unique properties that make them particularly effective in extracting valuable compounds from animals. Due to their adaptability, these systems offer distinct benefits compared to conventional organic solvents. They can be tailored on a large scale with minimal expense and low toxicity, high biodegradability, and biocompatibility.

Betaine/polyol-based DES was explored for the recovery of bioactive protein derivative-rich extracts contained in sardine processing waste streams [59]. While most DES does not significantly outperform water in total protein extraction yield, the lower polarity of the employed DES may facilitate the extraction of protein derivatives with a more hydrophobic character. These derivatives, known for their increased bioactive potential compared to hydrophilic counterparts, present a better outcome. Furthermore, observed synergistic or additive antimicrobial effects between solutes and DES suggest that the extracted compound could serve as a final product or formulation. This formulation holds promise as a functional ingredient or natural preservative across various industries, such as pharmaceuticals, nutraceuticals, or cosmetics, without the need for additional separation steps. Batista et al. [8] successfully extracted marine collagen from blue shark skins, through a new sustainable, faster, and more straightforward extraction approach using NADES composed of citric acid: xylitol: water when compared with the conventionally used extraction. The procedure proved highly efficient, eliminating the need for raw material pretreatment and substantially reducing extraction time from 96 h to just 1 h, consequently minimizing associated processing costs. Furthermore, the proposed methodology demonstrated a remarkable 2.5-fold enhancement in extraction yield compared to the conventional procedure, isolating over 21% of the protein content from the blue shark skin. DES facilitated the extraction of collagen with comparable properties in terms of purity, molecular weight, chemical composition, and cytotoxicity compared to organic solvents.

β-chitin from *Loligo vulgaris* squid *pens* was extracted for the first time with a simple DES solvent system, resulting in high purity and crystalline chitin with a yield of around 30% [47]. Among the DESs that were tested, the alkaline potassium carbonate:glycerol (KGLY) was the most efficient in obtaining purified β-chitin compared to the acidic and neutral DES systems. The selected β-chitin samples were then further characterized in terms of morphology, crystallinity, and thermal properties. The results showed that using KGLY led to high-quality β-chitin, similar to that obtained with the conventional alkaline extraction. Additionally, KGLY could be re-used three times without affecting the quality of the β-chitin. Because of the high purity and crystallinity of the obtained β-chitin, as well as the possibility of being recycled, the use of the alkaline KGLY DES was shown to be a particularly suitable method to isolate β-chitin and, potentially, for chitosan production.

Another interesting study was to obtain keratin from waste rabbit hair, where a novel, eco-friendly and benign choline chloride/oxalic acid deep eutectic solvent at a molar ratio of 1:2 was applied to dissolve the rabbit hair [71]. The keratin produced by dissolving in DES at 120 °C displayed flaky powders after freeze-drying and lower thermal stability than the raw rabbit hair and had a molecular weight ranging from 3.8 to 5.8 kDa. The sum of serine, glutamic acid, cysteine, leucine, and arginine accounted for 59.8 wt% of the total content of analyzed amino acids. The keratin obtained by dissolving rabbit hair in DES could be used as the dyeing and finishing agents for the chemical processing of textiles and may also find application in areas where low molecular weight keratin is required.

As discussed in the previous section, the integration of novel extraction techniques with DES leads to improved extraction outcomes, impacting not only the yield and biological activities but also the time and cost involved. Denga et al. [15] developed a green and efficient ultrasound-assisted natural deep eutectic solvent (NADES) method to recover astaxanthin (AXT) from shrimp shells. AXT was obtained through the utilization of six specific types of NADESs and acetone, with the extraction conditions being carefully optimized. Under the most favourable extraction conditions, the NADES formed by lysine-citric acid exhibited an extraction efficiency of 112.80 μg/g for AXT, surpassing that extracted with acetone by more than fivefold. Additionally, assessments of hydroxyl radical scavenging showed that AXT obtained with lysine-citric acid possessed a higher antioxidant capacity. It is evident that employing NADES combined with ultrasound extraction is not only economically viable but also offers environmental advantages in the extraction of bioactive substances from shrimp shell waste. The combined effects of an ultrasonic extraction method and DESs on the extraction of different bioactive substances from bee pollen have been reported [9]. Compared to the control group, which consisted of extracts obtained through the maceration method using ethanol as a solvent, the use of DESs led to higher values for both total individual amino acids and total individual organic acids. Furthermore, compounds such as myricetin, kaempferol, and quercetin were extracted at higher concentrations using DESs compared to controls. Also, the results obtained in antimicrobial activity tests showed that the DES groups had broad-spectrum antibacterial effects against all bacterial samples.

Other than ultrasonication, microwave-assisted extraction has also been employed with DESs. A time-saving, energy-efficient, and sustainable strategy consisting of microwave heating and deep eutectic solvent treatment is demonstrated to extract chitin from the crayfish shells [41]. Under the optimized extraction conditions, chitin is obtained with both a high yield (19.11%) and purity (97.44%). Additionally, the DES employed in the process can be effortlessly recycled and reused up to four times, contributing to a reduction in extraction costs and promoting sustainable development. Importantly, the extracted chitin can be further dissolved in DES to create a viscoelastic ink, showcasing superior rheological properties and 3D printability. This study introduces an environmentally friendly and uncomplicated approach to chitin production, offering a promising avenue for the high-value utilization of chitin as a 3D printing ink.

Despite the promising attributes of DES-based extraction methods, certain challenges persist. Most issues arise due to the high viscosity of DES, limiting the fluidity and ease of handling during the extraction process [81]. The selectivity of extraction is also a concern, as there may be limitations in precisely isolating the target compounds, leading to suboptimal results. Additionally, the back-extraction method, a crucial step in the recovery process, may suffer from inefficiencies. These challenges underscore the need for further refinement and optimization of DES-based extraction methodologies to enhance their practical applicability and overcome existing limitations in viscosity, selectivity, and back-extraction efficiency. Other than that, scaling up DES-based extraction methods for animal sources from laboratory to industrial scales may pose some challenges. The existing efficiency and cost-effectiveness demonstrated at the smaller scale may not seamlessly translate to larger operations. This shift demands meticulous consideration to ensure a successful upscaling process. The limited availability of comprehensive lab-scale results further underscores the need for extensive research and exploration.

Deep eutectic solvents exhibit versatile applications, showing potential in animal-derived sources. These environmentally friendly solvents offer scalability, low toxicity, and biocompatibility, presenting advantages over traditional solvents. Additionally, the integration of DES with novel extraction techniques, such as ultrasound-assisted and microwave-assisted extraction, enhances efficiency, reduces time, and promotes sustainability. However, challenges persist, primarily related to DES viscosity, extraction selectivity, and back-extraction efficiency. Overcoming these challenges through ongoing refinement is crucial for expanding the practical applicability of DES-based extraction methods, ensuring scalability, and realizing their full potential in the transition towards sustainable and environmentally friendly extraction practices on a larger scale.

6 Sustainable Aspects and Environmental Impact

6.1 The Sustainability of DES-Based Extraction in Comparison to Traditional Methods

The transition towards the use of Deep Eutectic Solvents (DES) in extraction methodologies marks a noteworthy leap forward in the realm of eco-friendly chemical practices. The inherent attributes of DES, including their capacity for biodegradation, minimal toxicity, and smaller ecological footprint, position them as a preferable choice over conventional solvent systems [43]. The eco-centric nature of DES-based extraction methods is highlighted by their formulation from components that are both accessible and derived from renewable sources, thereby contributing to a sustainable economic model and adhering to green chemistry's core principles [3].

The application scope of DES extends beyond mere extraction, finding use in the solvothermal generation of complex materials, such as ternary metal chalcogenide nanostructures, exemplified by wittichenite (Cu_3BiS_3), indicating the broad utility and adaptability of DES across varied scientific domains. Additionally, the creation of thermoelectric compounds like $AgBiS_2$ and Cu_3BiS_3 through microwave-assisted techniques further illustrates the significant role of DES in the advancement of material sciences and thermoelectric developments [3].

The principles of a circular economy and the transition to greener methodologies resonate well with the adoption of DES, highlighting a shift towards more sustainable resource management and production methodologies [54]. This approach is further validated by studies linking the economic valuation of natural ecosystems with sustainable economic frameworks, thereby underscoring the combined economic and environmental advantages of adopting green technologies such as DES [70].

From the perspective of renewable resource management, the adoption of DES is in harmony with the Renewable Substitutability (RS) strategy, which advocates for a gradual shift from finite resources to renewable alternatives, thus fostering a significant change in how resources are utilized [62]. Moreover, the European Union's ambitious environmental policies, backed by financial incentives from the European Commission, have played a crucial role in highlighting the multifaceted benefits—environmental, economic, and social—of a circular economic model, thereby encouraging the integration of sustainable practices like the use of DES [21].

In essence, the deployment of DES in extraction processes embodies a key innovation within the sphere of sustainable chemistry, aligning with green chemistry and circular economic principles. Its wide-ranging applicability in diverse synthesis processes and fields underscores its potential to catalyze the adoption of sustainable practices across various sectors.

6.2 Environmental Impact

The ecological footprint of extractions using Deep Eutectic Solvents (DES) presents a stark contrast to traditional methodologies dependent on volatile organic compounds (VOCs), demonstrating a substantial decrease in environmental detriment [68]. By curtailing VOC emissions, DES not only mitigates air contamination and health hazards in occupational settings but also diminishes risks associated with combustibility, thereby bolstering safety protocols in both research and industrial environments. Nonetheless, a holistic appraisal of DES, encompassing their creation, application, and eventual disposal, is imperative to grasp the full spectrum of their ecological consequences.

6.3 Potential Strategies for Minimizing Adverse Effects of DESs in Extraction of Active Compounds

Developing approaches to mitigate the negative impacts of Deep Eutectic Solvents (DESs) in extracting active ingredients is key to promoting eco-conscious and sustainable methodologies. Research has been directed towards crafting eco-friendly extraction mediums, underscoring the shift towards green technological solutions [40]. The deployment of DESs has been noted to significantly enhance the extraction yields of vital compounds, such as phenolic acids from diverse botanical sources, marking their efficacy in eco-efficient extraction endeavours [60]. Nonetheless, addressing the inherent drawbacks of DESs, including their potential toxicity, economic considerations, and their solid-state nature at ambient temperatures, remains imperative.

Adopting biodegradable and renewable materials in DES formulations stands out as a viable approach to diminish environmental repercussions and bolster sustainability measures [40]. Fine-tuning extraction parameters, such as temperature, duration, and the specific makeup of DES can further optimize yields while curtailing energy demands [9]. The incorporation of state-of-the-art methodologies like microwave-assisted and ultrasound-assisted extractions has proven to amplify efficiency and trim down operation times, fostering more eco-friendly extraction routines [9]. Additionally, executing thorough Life Cycle Assessments (LCAs) on DES-related procedures can unveil and attenuate possible ecological impacts across the solvent lifecycle, from its genesis to its eventual disposal [65]. The establishment of regulatory frameworks and standards governing the industrial application of DES can facilitate their broader acceptance and guarantee eco-responsible practices [17].

Explorations into the creation of DES-based functional materials for chromatographic separations have also been undertaken, revealing opportunities for greener and more efficient extraction modalities. The application of DESs in extracting bioactive constituents from natural reserves has proven their capacity to secure higher concentrations of such compounds, alongside exhibiting augmented biological effects, thereby underscoring their role in sustainable extraction paradigms [73]. Strategies aimed at reducing the detrimental effects of DESs in the extraction of active ingredients involve a holistic approach, spanning component selection, procedural optimization, detailed LCAs, and regulatory backing. These collective efforts are instrumental in advancing sustainable and eco-friendly extraction practices that resonate with the tenets of green chemistry and the circular economy.

7 Future Trends and Innovations

The field of green chemistry is undergoing significant transformation with the adoption of Deep Eutectic Solvents (DESs) for extracting bioactive compounds. This area is particularly noteworthy due to its potential impact on industries, such as pharmaceuticals, nutraceuticals, and food, where there is a growing demand for eco-friendly and efficient extraction methods. DESs are emerging as key players in this shift, offering novel approaches to bioactive compound extraction.

Investigations into the evolving landscape and novel approaches in utilizing Deep Eutectic Solvents (DESs) for bioactive compound extraction have yielded significant findings. The work of Potticary and colleagues in 2019 unveiled a new category of volatile DESs, showcasing their role in modulating pharmaceutical polymorphism and facilitating the ambient temperature creation of more effective versions of substances such as acetaminophen [57]. Pednekar and his team in 2013 accentuated the efficacy of DESs in the synthesis of 1,4-dihydropyridine derivatives through a one-pot, multi-component approach, underlining the importance of the solvents' reusability and compatibility with life, pivotal for eco-friendly chemical practices [55]. Furthermore, McCluskey and his group in 2018 [46] documented the inaugural occurrence of phospholipid monolayer formation in DESs, employing Bayesian modeling to decode the structure of these monolayers and the potential uses of the solvents. Together, these contributions highlight the dynamic and innovative uses of DESs in the field of sustainable chemistry, especially in the context of extracting and synthesizing bioactive molecules.

The ability to customize DESs is a promising development. Scientists are actively engaged in formulating DESs with precise physicochemical properties to target specific bioactive compounds. Adjusting the components of DESs allows for solvents that meet specific requirements regarding polarity, viscosity, and solubility [49]. This not only improves the extraction process but also lessens the need for additional purification steps.

The fusion of DESs with contemporary extraction technologies like supercritical fluid extraction, microwave-assisted extraction, and ultrasound-assisted extraction is becoming increasingly popular. These advanced methods, when used in conjunction with DESs, can significantly boost the efficiency of the extraction process. The synergy between DESs and these techniques enhances solvent penetration and cell wall disruption, facilitating the efficient release of bioactive components [61].

The use of DESs in extraction is expected to lean more towards automation and meticulous process optimization. Automated systems, capable of adjusting parameters in real-time, promise to fine-tune the extraction process for optimal results. The application of machine learning and artificial intelligence can offer predictions on the most effective DES compositions and optimal extraction conditions tailored to the specific bioactive compounds being targeted [63].

As DES-based extraction methods evolve, the focus on their environmental impact will intensify. Life cycle assessments evaluating the eco-friendliness of DESs from production through disposal will become commonplace. Innovations might include DESs made from bio-sourced or recycled materials, further boosting their environmental credentials. Additionally, strategies for the effective recycling and reuse of DESs will be essential in minimizing their ecological footprint [45].

The versatility of DESs is likely to lead to their use in a broader range of applications. Beyond traditional uses in plant extractions, DESs are expected to be employed in extracting substances from marine bioresources, microalgae, and waste biomass. This broadening of applications not only demonstrates the versatility of DESs but also contributes to waste reduction and promotes circular economic principles [23]. As the industrial application of DES-based extraction methods becomes more common, the creation of regulatory guidelines and standards will be critical. These measures will ensure the safe, effective, and eco-friendly use of these processes, aiding in their integration into various industrial sectors (Fig. 5).

The extraction of bioactive compounds using DESs is on the cusp of significant innovations and advancements. The development of customized DES formulations, combined with the integration of advanced extraction techniques, and a focus on automation and sustainability, is set to transform the extraction landscape. Ongoing

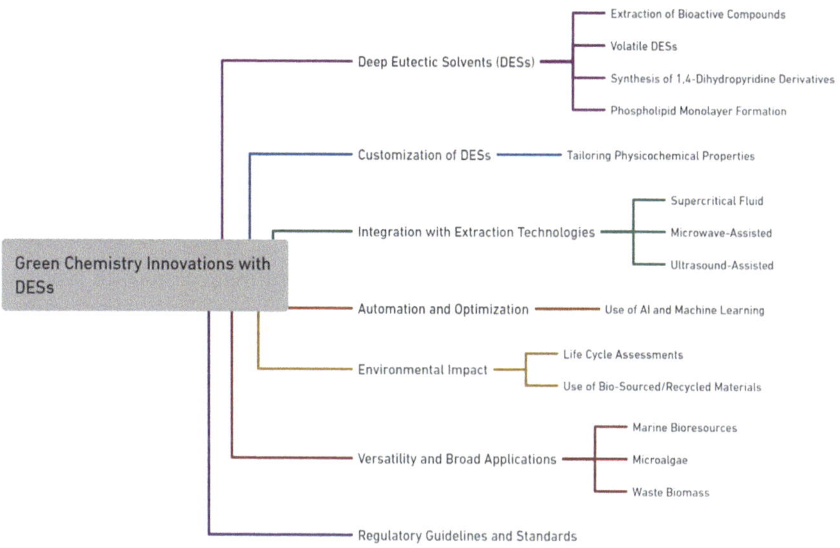

Fig. 5 Future trends an innovation in DESs extraction

research in this field will solidify the role of DES-based methods as a fundamental aspect of green chemistry, providing efficient, sustainable, and adaptable solutions for extracting bioactive compounds.

8 Conclusion

Investigating the role of Deep Eutectic Solvents (DESs) in isolating bioactive elements from flora and fauna marks a crucial transition towards methods that are both more eco-conscious and effective within green chemistry's domain. This segment has traced the progression of extraction methodologies, highlighting the shift from traditional solvents to the forward-thinking adoption of DESs, noted for their environmental compatibility, adaptability, and effectiveness in isolating a diverse range of bioactive elements. The progression from age-old extraction techniques to the introduction of DESs represents a deliberate stride towards reducing ecological footprints and augmenting the eco-friendliness of extraction operations. DESs stand out due to their distinctive eutectic characteristics and the feasibility of their derivation from components that are not only readily accessible and compatible with biological systems but also non-toxic, positioning them as a viable substitute for both water and conventional organic solvents. Their utilization spans beyond the extraction from plant sources to encompass precious compounds derived from animal sources, thereby illustrating the extensive utility and promise of DESs across multiple fields.

Moreover, this chapter has shed light on the challenges intrinsic to DESs, like their viscosity and the specificity of their extraction capabilities, which call for ongoing research and refinement to fully leverage their advantages. Additionally, the task of scaling DES-based techniques from the lab to industrial levels introduces added complexity, necessitating thorough comprehension and creative strategies to attain both efficiency and cost-effectiveness on a grander scale. Anticipated advancements and novel approaches in DES-based extraction aim to tackle these hurdles, concentrating on tailoring DES compositions for specific targets, merging DESs with cutting-edge extraction methods, and streamlining extraction processes through automation and optimization. The commitment to the ecological sustainability of DESs, assessed via thorough life cycle analyses, forms a fundamental aspect of forthcoming progress, ensuring that the integration of DESs aligns with the principles of green chemistry and the circular economy.

Employing DESs for the extraction of bioactive elements embodies a blend of efficiency, eco-friendliness, and innovation. As the exploration and development in this sphere progress, DESs are anticipated to assume a more prominent role within green chemistry, presenting practical alternatives to the hurdles posed by traditional extraction methods and charting a course towards extraction practices that are both more sustainable and environmentally benign. The evolution of DESs from a specialized to a widespread application in the extraction of bioactive substances not only underscores the creativity inherent in green chemistry but also serves as a harbinger of forthcoming sustainable technological breakthroughs.

Acknowledgements We are grateful for the research funding from Sultan Sharif Ali Islamic University to HMS through the internal research grant scheme no. [UNISSA/PPP/Grant/CRP0012—2023/2024].

References

1. Abbott, A.P., Capper, G., Davies, D.L., Rasheed, R.K., Tambyrajah, V.: Novel solvent properties of choline chloride/urea mixtures. Chem. Commun. **1**, 70–71 (2003). https://doi.org/10.1039/b210714g
2. Abduljabar, M.K., Kasim, K.F., Ma'Radzi, A.H., Seng, N.S.S.: Recent advances in extraction of Clinacanthus nutans: a review. J. Phys.: Conf. Ser. **1424**(1) (2019). https://doi.org/10.1088/1742-6596/1424/1/012001
3. Adeyemi, A.N., Clemente, M., Lee, S., Mantravadi, A., Zaikina, J.V.: Deep eutectic solvent-assisted microwave synthesis of thermoelectric $AgBiS_2$ and Cu_3BiS_3.. ACS Appl. Energy Mater. (2022). https://doi.org/10.1021/acsaem.2c02336
4. Almohasin, J.A., Balag, J., Miral, V.G., Moreno, R.V., Tongco, L.J., Lopez, E.C.R.: Green Solvents for Liquid–Liquid Extraction: Recent Advances and Future Trends, pp. 1–11 (2023)
5. Amoroso, R., Hollmann, F., Maccallini, C.: Choline chloride-based des as solvents/catalysts/chemical donors in pharmaceutical synthesis. Molecules **26**(20) (2021). https://doi.org/10.3390/molecules26206286
6. Basaiahgari, A., Panda, S., Gardas, R.L.: Effect of ethylene, diethylene, and triethylene glycols and glycerol on the physicochemical properties and phase behavior of benzyltrimethyl and

benzyltributylammonium chloride based deep eutectic solvents at 283.15–343.15 K. J. Chem. Eng. Data **63**(7), 2613–2627 (2018). https://doi.org/10.1021/acs.jced.8b00213

7. Bashir, I., Dar, A.H., Dash, K.K., Pandey, V.K., Fayaz, U., Shams, R., Srivastava, S., Singh, R.: Deep eutectic solvents for extraction of functional components from plant-based products: a promising approach. Sustain. Chem. Pharm. **33**, 101102 (2023). https://doi.org/10.1016/J.SCP.2023.101102
8. Batista, M.P., Fernández, N., Gaspar, F.B., Bronze, M.d.R., Duarte, A.R.C.: Extraction of biocompatible collagen from blue shark skins through the conventional extraction process intensification using natural deep eutectic solvents. Front. Chem. **10**, 937036 (2022). https://doi.org/10.3389/FCHEM.2022.937036/BIBTEX
9. Çelik, S., Kutlu, N., Gerçek, Y.C., Bayram, S., Pandiselvam, R., Bayram, N.E.: Optimization of ultrasonic extraction of nutraceutical and pharmaceutical compounds from bee pollen with deep eutectic solvents using response surface methodology. Foods **11**(22), 3652 (2022). https://doi.org/10.3390/FOODS11223652
10. Chaves, J.O., de Souza, M.C., da Silva, L.C., Lachos-Perez, D., Torres-Mayanga, P.C., Machado, A.P.d.F., Forster-Carneiro, T., Vázquez-Espinosa, M., González-de-Peredo, A.V., Barbero, G.F., Rostagno, M.A.: Extraction of flavonoids from natural sources using modern techniques. Front. Chem. **8**(September) (2020). https://doi.org/10.3389/fchem.2020.507887
11. Chen, J., Liu, M., Wang, Q., Du, H., Zhang, L.: Deep eutectic solvent-based microwave-assisted method for extraction of hydrophilic and hydrophobic components from radix Salviae miltiorrhizae. Molecules **21**(10) (2016). https://doi.org/10.3390/molecules21101383
12. Christen, P., Kaufmann, B.: Recent extraction techniques for natural products: microwave-assisted extraction and pressurised solvent extraction. Phytochem. Anal. **113**(13), 105–113 (2002)
13. Cortes-Clerget, M., Yu, J., Kincaid, J.R.A., Walde, P., Gallou, F., Lipshutz, B.H.: Water as the reaction medium in organic chemistry: from our worst enemy to our best friend. Chem. Sci. **12**(12), 4237–4266 (2021). https://doi.org/10.1039/d0sc06000c
14. Dardavilla, M.M., Pappou, S., Savvidou, Maria. G., Louli, V., Katapodis, P., Stamatis, H., Magoulas, K., Voutsas, E.: Extraction of bioactive compounds from C. vulgaris biomass using deep eutectic solvents. Molecules **28**(415) (2023)
15. Denga, M., Qu, Y., Wu, T., Na, Y., Liang, N., Zhao, L.: Amino acid-based natural deep eutectic solvent combined with ultrasonic extraction: green extraction of astaxanthin from shrimp shells. Biomass Convers. Biorefinery 1–10 (2023). https://doi.org/10.1007/S13399-023-04660-5/TABLES/3
16. Doldolova, K., Bener, M., Lalikoğlu, M., Aşçı, Y.S., Arat, R., Apak, R.: Optimization and modeling of microwave-assisted extraction of curcumin and antioxidant compounds from turmeric by using natural deep eutectic solvents. Food Chem. **353**, 129337 (2021). https://doi.org/10.1016/J.FOODCHEM.2021.129337
17. Duan, L., Li, D., Guo, L., Li, P., Liu, E.: Comprehensive evaluation of deep eutectic solvents in extraction of bioactive natural products. ACS Sustain. Chem. Eng. (2016). https://doi.org/10.1021/acssuschemeng.6b00091
18. Dzobo, K.: The role of natural products as sources of therapeutic agents for innovative drug discovery. Compr. Pharmacol. **2**(January), 408–422 (2022). https://doi.org/10.1016/B978-0-12-820472-6.00041-4
19. El Achkar, T., Fourmentin, S., Greige-Gerges, H.: Deep eutectic solvents: an overview on their interactions with water and biochemical compounds. J. Mol. Liq. **288** (2019). https://doi.org/10.1016/j.molliq.2019.111028
20. Elgharbawy, A., Syed Putra, S., Khan, H., Azmi, N., Sani, M., Ab llah, N., Hayyan, A., Jewaratnam, J., Basirun, W.: Menthol and fatty acid-based hydrophobic deep eutectic solvents as media for enzyme activation. Processes **11**(2), 547 (2023). https://doi.org/10.3390/pr11020547
21. Fric, U.: Impact of circular economy as the EU's ambitious policy. Res. Soc. Chang. (2019). https://doi.org/10.2478/rsc-2019-0010

22. Garcia-Vaquero, M., Rajauria, G., Tiwari, B.: Conventional extraction techniques: solvent extraction. In: Sustainable Seaweed Technologies: Cultivation, Biorefinery, and Applications. Elsevier Inc. (2020). https://doi.org/10.1016/B978-0-12-817943-7.00006-8
23. Gaston, K.J., Anderson, K., Shutler, J.D., Brewin, R.J.W., Yan, X.: Environmental impacts of increasing numbers of artificial space objects. Front. Ecol. Environ. (2023). https://doi.org/10.1002/fee.2624
24. Gilmore, M., Swadzba-Kwasny, M., Holbrey, J.D.: Thermal properties of choline chloride/urea system studied under moisture-free atmosphere. J. Chem. Eng. Data **64**(12), 5248–5255 (2019). https://doi.org/10.1021/acs.jced.9b00474
25. Gomez, F.J.V., Espino, M., de los Angeles Fernandez, M., Raba, J., Silva, M.F.: Enhanced electrochemical detection of quercetin by natural deep eutectic solvents. Analytica Chimica Acta, **936**, 91–96 (2016). https://doi.org/10.1016/j.aca.2016.07.022
26. Gonçalves, A.R.P., Paredes, X., Cristino, A.F., Santos, F.J.V., Queirós, C.S.G.P.: Ionic liquids—a review of their toxicity to living organisms. Int. J. Mol. Sci. **22**(11) (2021). https://doi.org/10.3390/ijms22115612
27. Gu, J., Lin, L., Zhao, M.: Demonstration of feasibility and effectiveness of deep eutectic solvent-water system extraction of RG-I type pectin from wolfberry based on target polysaccharide, solvent and their interactions. Food Hydrocoll. **144**, 109027 (2023). https://doi.org/10.1016/J.FOODHYD.2023.109027
28. Guo, W., Hou, Y., Wu, W., Ren, S., Tian, S., Marsh, K.N.: Separation of phenol from model oils with quaternary ammonium salts via forming deep eutectic solvents. Green Chem. **15**(1), 226–229 (2013). https://doi.org/10.1039/c2gc36602a
29. Habibi, E., Ghanemi, K., Fallah-Mehrjardi, M., Dadolahi-Sohrab, A.: A novel digestion method based on a choline chloride-oxalic acid deep eutectic solvent for determining Cu, Fe, and Zn in fish samples. Anal. Chim. Acta **762**, 61–67 (2013). https://doi.org/10.1016/j.aca.2012.11.054
30. Hansen, B.B., Spittle, S., Chen, B., Poe, D., Zhang, Y., Klein, J.M., Horton, A., Adhikari, L., Zelovich, T., Doherty, B.W., Gurkan, B., Maginn, E.J., Ragauskas, A., Dadmun, M., Zawodzinski, T.A., Baker, G.A., Tuckerman, M.E., Savinell, R.F., Sangoro, J.R.: Deep eutectic solvents: a review of fundamentals and applications. Chem. Rev. **121**(3), 1232–1285 (2021). https://doi.org/10.1021/acs.chemrev.0c00385
31. Ivanović, M., Razboršek, M.I., Kolar, M.: Innovative extraction techniques using deep eutectic solvents and analytical methods for the isolation and characterization of natural bioactive compounds from plant material. Plants **9**(11), 1–29 (2020). https://doi.org/10.3390/plants9111428
32. Jakovljević, M., Vladić, J., Vidović, S., Pastor, K., Jokić, S., Molnar, M., Jerković, I.: Application of deep eutectic solvents for the extraction of rutin and rosmarinic acid from Satureja montana L. and evaluation of the extracts antiradical activity. Plants **9**(2), 153 (2020). https://doi.org/10.3390/PLANTS9020153
33. Khan, H.W., Elgharbawy, A.A.M., Bustam, M.A., Goto, M., Moniruzzaman, M.: Vegetable oil-ionic liquid-based emulsion liquid membrane for the removal of lactic acid from aqueous streams: emulsion size, membrane breakage, and stability study. ACS Omega **7**(36), 32176–32183 (2022)
34. Khan, H.W., Reddy, A.V.B., Bustam, M.A., Goto, M., Moniruzzaman, M.: Development and optimization of ionic liquid-based emulsion liquid membrane process for efficient recovery of lactic acid from aqueous streams. Biochem. Eng. J. **176**, 108216 (2021). https://doi.org/10.1016/j.bej.2021.108216
35. Khan, H.W., Zailan, A.A., Bhaskar Reddy, A.V., Goto, M., Moniruzzaman, M.: Ionic liquid-based dispersive liquid–liquid microextraction of succinic acid from aqueous streams: COSMO-RS screening and experimental verification. Environ. Technol. (United Kingdom) (2023). https://doi.org/10.1080/09593330.2023.2234669
36. Khan, P.A., Johl, S.K., Akhtar, S.: Vinculum of sustainable development goal practices and firms' financial performance: a moderation role of green innovation. J. Risk Financ. Manag. **15**(3), 96 (2022). https://doi.org/10.3390/jrfm15030096

37. Khan, P.A., Johl, S.K., Johl, S.K.: Does adoption of ISO 56002–2019 and green innovation reporting enhance the firm sustainable development goal performance? An emerging paradigm. Bus. Strateg. Environ. **30**(7), 2922–2936 (2021). https://doi.org/10.1002/bse.2779
38. Khezeli, T., Daneshfar, A.: Synthesis and application of magnetic deep eutectic solvents: novel solvents for ultrasound assisted liquid-liquid microextraction of thiophene. Ultrason. Sonochem. **38**, 590–597 (2017). https://doi.org/10.1016/j.ultsonch.2016.08.023
39. Koraqi, H., Qazimi, B., Çesko, C., Petkoska, A.T.: Environmentally friendly extraction of bioactive compounds from Rosa canina L. fruits using deep eutectic solvent (DES) as green extraction media. Acta Chimica Slovenica, **69**(3), 665–673 (2022). https://doi.org/10.17344/acsi.2022.7559
40. Li, L., Liu, Y., Wang, Z., Yang, L., Liu, H.: Development and applications of deep eutectic solvent derived functional materials in chromatographic separation. J. Sep. Sci. (2020). https://doi.org/10.1002/jssc.202000523
41. Li, Z., Li, M.C., Liu, C., Liu, X., Lu, Y., Zhou, G., Liu, C., Mei, C.: Microwave-assisted deep eutectic solvent extraction of chitin from crayfish shell wastes for 3D printable inks. Ind. Crops Prod. **194**, 116325 (2023). https://doi.org/10.1016/J.INDCROP.2023.116325
42. Lomba, L., Ribate, M.P., Sangüesa, E., Concha, J., Garralaga, M.P., Errazquin, D., Garc, C.B., Giner, B.: Deep eutectic solvents: are they safe? Appl. Sci. **11**(10061), 1–21 (2021)
43. Mahto, B., Khan, A.A., Barhoi, A., Hussain, S.: Hierarchical Cu_3BiS_3 nanostructures with thermally controlled morphology for photoconductive, photothermal, and catalytic applications. ACS Appl. Nano Mater. (2023). https://doi.org/10.1021/acsanm.3c00496
44. Martins, M.A.R., Pinho, S.P., Coutinho, J.A.P.: Insights into the nature of eutectic and deep eutectic mixtures. J. Solution Chem. **48**(7), 962–982 (2019). https://doi.org/10.1007/s10953-018-0793-1
45. Matthews, L., Ruscigno, S., Rogers, S.E., Bartlett, P., Johnson, A.J., Sochon, R., Briscoe, W.H.: Fracto-eutectogels: SDS fractal dendrites via counterion condensation in a deep eutectic solvent. Phys. Chem. Chem. Phys. (2021). https://doi.org/10.1039/d1cp01370j
46. McCluskey, A.R., Sanchez-Fernandez, A., Edler, K.J., Parker, S.C., Jackson, A.J., Campbell, R.A., Arnold, T.: Bayesian determination of the effect of a deep eutectic solvent on the structure of lipid monolayers. Phys. Chem. Chem. Phys. **21**(11), 6133–6141 (2019)
47. Mcreynolds, C., Adrien, A., Brugerolle De Fraissinette, N., Olza, S., Fernandes, S.: Deep eutectic solvents for the extraction of β-chitin from Loligo vulgaris squid pens: a sustainable way to valorize fishery by-products. Biomass Convers. Biorefinery (2023). https://doi.org/10.1007/s13399
48. Migliorati, V., Sessa, F., D'Angelo, P.: Deep eutectic solvents: a structural point of view on the role of the cation. Chem. Phys. Lett.: X **2**, 100001 (2019). https://doi.org/10.1016/j.cpletx.2018.100001
49. Molnar, M., Kraljević, T.G., Pavić, V., Rastija, V., Komar, M.: Environmentally friendly approach to the synthesis of 3-[benzylideneamino]-2-methylquinazolin-4(3*H*)-one derivatives and calculation of their toxicity. Chem. Biodivers. (2023). https://doi.org/10.1002/cbdv.202300575
50. Mussarat, S., Ali, R., Ali, S., Mothana, R.A., Ullah, R., Adnan, M.: Medicinal animals and plants as alternative and complementary medicine in southern regions of Khyber Pakhtunkhwa, Pakistan. Front. Pharmacol. **12**(August), 1–17 (2021). https://doi.org/10.3389/fphar.2021.649046
51. Nam, M.W., Zhao, J., Lee, M.S., Jeong, J.H., Lee, J.: Enhanced extraction of bioactive natural products using tailor-made deep eutectic solvents: application to flavonoid extraction from Flos sophorae. Green Chem. **17**(3), 1718–1727 (2015)
52. Nanda, B., Sailaja, M., Mohapatra, P., Pradhan, R.K., Nanda, B.B.: Green solvents: a suitable alternative for sustainable chemistry. Mater. Today: Proc. **47**, 1234–1240 (2021). https://doi.org/10.1016/j.matpr.2021.06.458
53. Nekkaa, A., Benaissa, A., El Djalil Lalaouna, A., Dupire, F., Risler, A., Mutelet, F., Canabady-Rochelle, L.: Green and innovative extraction of polyphenols from Rhamnus alaternus using natural deep eutectic solvents and evaluation of their bioactivities. J. Appl. Res. Med. Aromat. Plants **35**, 100503 (2023). https://doi.org/10.1016/J.JARMAP.2023.100503

54. O'Born, R.J., Heimdal, A.: Experiences from Teaching Circular Economy Concepts to Engineering Students (2022). https://doi.org/10.35199/epde.2022.76
55. Pednekar, S., Bhalerao, R., Ghadge, N.: One-pot multi-component synthesis of 1,4-dihydropyridine derivatives in biocompatible deep eutectic solvents. J. Chem. Sci. **125**, 615–621 (2013)
56. Płotka-Wasylka, J., de la Guardia, M., Andruch, V., Vilková, M.: Deep eutectic solvents vs ionic liquids: similarities and differences. Microchem. J. **159**(July) (2020). https://doi.org/10.1016/j.microc.2020.105539
57. Potticary, J., Hall, C., Hamilton, V., McCabe, J.F., Hall, S.R.: Crystallization from volatile deep eutectic solvents. Cryst. Growth Des. **20**(5), 2877–2884 (2020)
58. Quitério, E., Grosso, C., Ferraz, R., Delerue-Matos, C., Soares, C.: A critical comparison of the advanced extraction techniques applied to obtain health-promoting compounds from seaweeds. Mar. Drugs **20**(11), 1–40 (2022). https://doi.org/10.3390/md20110677
59. Rodrigues, L.A., Leonardo, I.C., Gaspar, F.B., Roseiro, L.C., Duarte, A.R.C., Matias, A.A., Paiva, A.: Unveiling the potential of betaine/polyol-based deep eutectic systems for the recovery of bioactive protein derivative-rich extracts from sardine processing residues. Sep. Purif. Technol. **276**(May) (2021). https://doi.org/10.1016/j.seppur.2021.119267
60. Rodríguez-Martínez, B., Ferreira-Santos, P., Alfonso, I.M., Martínez, S., Genisheva, Z., Gullón, B.: Deep eutectic solvents as a green tool for the extraction of bioactive phenolic compounds from avocado peels. Molecules (2022). https://doi.org/10.3390/molecules27196646
61. Sivagourounadin, K., Ravichandran, M., Rajendran, P.: National guidelines for gene therapy product (2019): a road-map to gene therapy products development and clinical trials. Perspect. Clin. Res. (2021). https://doi.org/10.4103/picr.picr_189_20
62. Srinivasan, R.S., Campbell, D.E.: Renewable substitutability index: maximizing renewable resource use in buildings. Buildings (2015). https://doi.org/10.3390/buildings5020581
63. Srivastava, S.: Knoevenagel condensation and Michael addition in bio-renewable deep eutectic solvent: facile synthesis of a library of bis-enol derivatives. ChemistrySelect (2020). https://doi.org/10.1002/slct.201904806
64. Tang, B., Bi, W., Zhang, H., Row, K.H.: Deep eutectic solvent-based HS-SME coupled with GC for the analysis of bioactive terpenoids in Chamaecyparis obtusa leaves. Chromatographia **77**(3–4), 373–377 (2014). https://doi.org/10.1007/s10337-013-2607-3
65. Tian, H., Wang, J., Li, Y., Bi, W., Chen, D.D.Y.: Recovery of natural products from deep eutectic solvents by mimicking denaturation. ACS Sustain. Chem. Eng. (2019). https://doi.org/10.1021/acssuschemeng.9b01012
66. Tien, N.N.T., Le, N.L., Khoi, T.T., Richel, A.: Optimization of microwave-ultrasound-assisted extraction (MUAE) of pectin from dragon fruit peels using natural deep eutectic solvents (NADES). J. Food Process. Preserv. **46**(1), e16117 (2022)
67. Tomé, L.I.N., Baião, V., da Silva, W., Brett, C.M.A.: Deep eutectic solvents for the production and application of new materials. Appl. Mater. Today **10**, 30–50 (2018). https://doi.org/10.1016/j.apmt.2017.11.005
68. Vinci, G., Maddaloni, L., Prencipe, S.A., Orlandini, E., Sambucci, M.: Simple and reliable eco-extraction of bioactive compounds from dark chocolate by deep eutectic solvents. A sustainable study. Int. J. Food Sci. Technol. (2023). https://doi.org/10.1111/ijfs.16315
69. Vo, T.P., Pham, T.V., Tran, T.N.H., Vo, L.T.V., Vu, T.T., Pham, N.D., Nguyen, D.Q.: Ultrasonic-assisted and microwave-assisted extraction of phenolics and terpenoids from Abelmoschus sagittifolius (Kurz) Merr roots using natural deep eutectic solvents. ACS Omega **8**(32), 29704–29716 (2023). https://doi.org/10.1021/ACSOMEGA.3C03929/ASSET/IMAGES/LARGE/AO3C03929_0005.JPEG
70. Wahyuningrum, D., Gravitiani, E., Sartika, R.: The interrelationship of sustainable economic value of watersheds using contingent valuation method approach with circular economy: a literature study. In: IOP Conference Series Earth and Environmental Science (2021). https://doi.org/10.1088/1755-1315/940/1/012036
71. Wang, D., Yang, X.H., Tang, R.C., Yao, F.: Extraction of keratin from rabbit hair by a deep eutectic solvent and its characterization. Polymers **10**(9), 993 (2018). https://doi.org/10.3390/POLYM10090993

72. Wang, H., Hu, L., Liu, X., Yin, S., Lu, R., Zhang, S., Zhou, W., Gao, H.: Deep eutectic solvent-based ultrasound-assisted dispersive liquid-liquid microextraction coupled with high-performance liquid chromatography for the determination of ultraviolet filters in water samples. J. Chromatogr. A **1516**, 1–8 (2017). https://doi.org/10.1016/j.chroma.2017.07.073
73. Wang, R., He, R., Li, Z., Wang, L.: LC-Q-Orbitrap-MS/MS characterization, antioxidant activity, and A-glucosidase-inhibiting activity with in silico analysis of extract from Clausena indica (Datz.) Oliv fruit pericarps. Front. Nutr. (2021). https://doi.org/10.3389/fnut.2021.727087
74. Warsi Khan, H., Kaif Khan, M., Moniruzzaman, M., Al Mesfer, M.K., Danish, M., Irshad, K., Yusuf, M., Kamyab, H., Chelliapan, S.: Evaluating ionic liquids for its potential as eco-friendly solvents for naproxen removal from water sources using COSMO-RS: computational and experimental validation. Environ. Res. 231 (2023). https://doi.org/10.1016/j.envres.2023.116058
75. Wen, L., Fan, C., Zhao, X., Cao, X.: Rapid extraction of bioactive compounds from gardenia fruit using new and recyclable deep eutectic solvents. J. Sep. Sci. **46**(17), 2300163 (2023). https://doi.org/10.1002/JSSC.202300163
76. Xu, Y., Wang, Q., Hou, Y.: Efficient purification of R-phycoerythrin from marine algae (Porphyra yezoensis) based on a deep eutectic solvents aqueous two-phase system. Mar. Drugs **18**(12), 1–14 (2020). https://doi.org/10.3390/md18120618
77. Xu, Z., Cai, Y., Ma, Q., Zhao, Z., Yang, D., Xu, X.: Optimization of extraction of bioactive compounds from Baphicacanthus cusia leaves by hydrophobic deep eutectic solvents. Molecules **26**(6) (2021). https://doi.org/10.3390/molecules26061729
78. Yang, Z.: Toxicity and biodegradability of deep eutectic solvents and natural deep eutectic solvents. In: Deep Eutectic Solvents: Synthesis, Properties, and Applications, pp. 43–60 (2019). https://doi.org/10.1002/9783527818488.ch3
79. Zainal-Abidin, M.H., Hayyaṅ, M., Hayyan, A., Jayakumar, N.S.: New horizons in the extraction of bioactive compounds using deep eutectic solvents: a review. Anal. Chim. Acta **979**, 1–23 (2017). https://doi.org/10.1016/j.aca.2017.05.012
80. Zhang, Q.W., Lin, L.G., Ye, W.C.: Techniques for extraction and isolation of natural products: a comprehensive review. Chin. Med. (United Kingdom) **13**(1), 1–26 (2018). https://doi.org/10.1186/s13020-018-0177-x
81. Zhou, Y., Wu, W., Zhang, N., Soladoye, O.P., Zhang, Y., Fu, Y.: Deep eutectic solvents as new media for green extraction of food proteins: opportunity and challenges. Food Res. Int. **161**, 111842 (2022). https://doi.org/10.1016/J.FOODRES.2022.111842

Extraction of Protein from Plant Materials Using Deep Eutectic Solvents

Noorfatimah Yahaya, Nur Nadhirah Mohamad Zain, Sazlinda Kamaruzaman, Ahmad Husaini Mohamed, and Kartika A. Madurani

Abstract Over the years, organic and aqueous solvents in both conventional and emerging techniques have been employed for protein extraction. Issues such as protein denaturation and alteration, environmental pollution, and toxicity have prompted further research. In the past few years, deep eutectic solvents (DES) have emerged as innovative solvents for protein extraction, composed of hydrogen bond donor (HBD) and hydrogen bond acceptor (HBA) molecules. Due to their excellent physicochemical properties, relatively low cost, and lower toxicity, DES have been identified as sustainable solvents for targeted molecules. This chapter reviews the application of DES for extracting protein from plant materials from 2019 to 2024. The chapter described the advantages and disadvantages, as well as the mechanisms and factors affecting the extraction efficiency of proteins. The discussion also delved into the limitations and challenges to bridge the gap between innovations and practical implementations.

N. Yahaya (✉) · N. N. M. Zain
Department of Toxicology, Advanced Medical and Dental Institute (AMDI), Universiti Sains Malaysia, Bertam, Kepala Batas, Penang, Malaysia
e-mail: noorfatimah@usm.my

S. Kamaruzaman
Department of Chemistry, Faculty of Science, Universiti Putra Malaysia, UPM Serdang, Seri Kembangan, Selangor Darul Ehsan, Malaysia

A. H. Mohamed
School of Chemistry and Environment, Universiti Teknologi MARA (UiTM) Negeri Sembilan, Kuala Pilah, Negeri Sembilan, Malaysia

K. A. Madurani
Chemistry Department, Faculty of Science and Data Analytics, Institut Teknologi Sepuluh Nopember, Surabaya, Indonesia

D. N. Jimat (ed.), *Deep Eutectic Solvents*, Engineering Materials,
https://doi.org/10.1007/978-981-95-1912-5_6

1 Introduction

Protein is an essential macronutrient for maintaining the growth and development of the body. However, the situation will worsen if the supply and dissemination of protein-rich foods and food supplements are not sustained, if mitigation plans for climate change are not implemented, and if renewable plant-based foods are not utilized as a source of energy and nutrition [1]. The primary roles of proteins are connected with their nutritional, physiological, and chemical composition [2]. People mainly depend on animal proteins such as meat, fish, dairy, and others of animal origin in their daily diet, however, the supply is relatively expensive due to the limited availability of these resources and the outcomes of climate change, freshwater level depletion, loss of biodiversity and detection of several human diseases [1, 3]. Compared to animal protein, plant protein consumption, and utilization have gained much attention [4]. Plant proteins are considered more ethically sound compared to animal proteins, and animal welfare associations are becoming increasingly concerned about meat proteins [5]. The increased awareness of the importance of high-quality protein in the diet has stimulated researchers to develop sustainable protein sources [1]. However, their extraction is usually performed with inefficient and relatively expensive methods that require the use of toxic solvents/reagents, which can severely limit their applicability [6].

In general, protein consists of soluble and insoluble fractions. Protein hydrolysates are a group of bio-stimulants consisting of polypeptides and amino acids. They are synthesized from soluble protein through the hydrolysis process. Insoluble proteins, however, cannot be dissolved, and they require mechanical disruption, enzyme treatment, chemical treatment, pH adjustment, and sequential solvent extraction for an efficient extraction process [7]. Protein from plants can be used as foaming agents, gelling agents, and emulsifiers in food systems [8]. In addition, they have been shown to have therapeutic benefits. For examples, metabolic disorders such as cancer and diabetes can be prevented by consuming plant proteins [9]. The functions and therapeutic properties are influenced mainly by their source and conformation, which is dependent on the protein isolation techniques and processing conditions, including pH and temperature [10].

Various dry and wet protein extraction (subcritical water extraction, reverse micelles extraction and aqueous two-phase systems extraction) have been reported previously [3]. A comprehensive review of conventional methods, such as aqua-based, salt, solvent, detergent, alkali, and non-conventional cell-disruptive methods, such as enzymes, microwave, high pressure, pulse-field, homogenization, and sonication, has been rarely discussed. Conventional methods can be referred to as common methods that sometimes result in lower extraction yield due to protein degradation. This decrease in protein yield is influenced by extraction time, type of solvents, pH, and temperature. Thus, current research trends are more focused on non-thermal green technologies to enhance extraction efficiency and reduce protein degradation during the extraction process [1]. In reviewing the existing literature [1, 11, 12], type of solvent was identified as one of the significant variables in the extraction of protein from plants.

The extraction of protein is carried out using common solvents such as water, alkali, acids, and organic solvents [13]. Acidic solvents are preferable due to less cell wall degradation, while alkaline solvents could result in protein denaturation and alteration, making them unsuitable for the purpose [1, 14]. On the other hands, recovered protein in organic solvents is not suitable for human consumption [15]. Further, combining water with enzymes, water at sub-critical conditions and reverse micelles have also been proposed for the extraction of protein. Yet, the process is considered less cost-effective [16]. Although water is the preferred greenest choice, most of macromolecules and organic compounds require the use of less polar solvent. In addition, carbon dioxide used in the supercritical fluids are expensive and show low polarity and is not being universally useful [17]. Therefore, scientists and researchers are moving in the direction of developing a more environmentally friendly and sustainable solvent approach as an alternative to conventional methods.

Deep eutectic solvents (DES) have recently been introduced as an environmentally friendly solvent [18]. DES are eutectic mixture of two or more ionic/non-ionic compounds in a specific molar ratio where external force such as heating, stirring, mechanical forces, sonication, or microwave is required to lower their individual melting point to a common eutectic point [12]. DES are in demand because they are relatively inexpensive, easy to synthesize, non-toxic, biodegradable, and sustainable [12, 19]. DES are called natural deep eutectic solvent (NADES) when their components are natural primary metabolites such as organic acids, amino acids, sugars, urea, etc. [17]. NADES are more environmentally friendly than classical DES and many of them have been reported previously [20]. Therefore, in this chapter, we present an overview of the application of DES for the extraction of protein from plant materials. The optimum conditions of the reviewed methodologies are discussed, and tables containing the applications of these DES since 2019 are presented.

2 Application of DES for Protein Extraction/Purification from Plant Materials

Researchers have moved towards refining extraction methods to improve current methodologies and simplify the extraction of proteins. Thus, this section will focus on the DES-based extraction/purification of protein from plant materials. The discussion comprises the analysis of developed extraction methods, optimum extraction conditions, and a summary of findings that include protein yields. A comprehensive literature review based on the years 2019 to 2024 pertaining to the developed DES-based extraction and purification techniques is presented in Table 1.

Li and co-workers [21] conducted a comprehensive study involving the extraction methodology. Initially, the kelp underwent a process of cleaning, drying at 50 °C, and subsequent grinding into a powder. Subsequently, 200 mg of the kelp

Table 1 Applications of DES for extraction and purification of proteins from plant materials

Type of plant materials	Type of extraction/ purification technique (s)	Composition and molar ratio of DES	Conditions of extraction	Summary of findings	Year	Reference
Kelp	Air Assisted Dispersive Liquid–Liquid Microextraction (AA-DLLME)	Choline chloride chloride (HBA): Caffeic acid Glycerol (HBDs) molar ratio: 1:1:1	Extraction temp—25 °C Extraction time—10 min Solid: liquid—200 mg of the kelp with 4 mL of the DES Stirring rate—150 rpm	The isolated concentrations of D-(+)-galactose, L-(-)-fucose, DL-tyrosine, and DL-valine in kelp were 16.7 ± 0.2 $mg.g^{-1}$, 8.6 ± 0.2 $mg.g^{-1}$, 2.6 ± 0.1 $mg.g^{-1}$, and 1.6 ± 0.1 $mg.g^{-1}$, respectively	2019	[21]
Kelp	Liquid extraction (Aqueous biphasic systems (ABSs)	Amino acids and polyols ([amino acids][polyols]) Tetrabutylammonium chloride and polypropylene glycol	Extraction temp—25 °C Extraction time—3 h Solid: liquid—50 g shredded fresh porcine pancreas tissue and 150 mL of the DES Stirring rate—7000 rpm	97.30% chymotrypsin tended to distribute into the [Pro][Xyl]-rich phase under optimum conditions	2019	[22]
Bioactive Compounds from *Mentha arvensis*	Ultrasound-assisted extraction (UAE)	Choline chloride (HBA): Glycerol (HBD) molar ratio: 1:2	Extraction temp—70 °C Extraction time—60 min Solid: liquid—2 g dried plant material with 20 mL of DESs	The total phenolic contents (TPCs) - 117 mg GAE/g Total flavonoids contents (TFCs) - 80 mg QE/g 2, 2-diphenyl-1-picrylhydrazyl (DPPH) radical inhibition - 94%	2019	[25]

(continued)

Table 1 (continued)

Type of plant materials	Type of extraction/ purification technique (s)	Composition and molar ratio of DES	Conditions of extraction	Summary of findings	Year	Reference
Pomegranate peel	Liquid extraction	Choline chloride (HBA): Urea or Ethylene glycol or Glycerol or Acetic acid or Glucose or Sorbitol or Citric acid (HBDs) All DES contained water (3rd component) to control solvent viscosity molar ratio: 1:1:10	Extraction temp—120 °C Extraction time—3 min Solid: liquid—150 mg of dried pomegranate peels with 5 mL of the DES Stirring rate (by centrifugation)—10 min at 4000 × g	The extraction yield by using DES is 20 ± 1 mg protein/g peel This is more than twice the amount of proteins extracted by PLE (9 ± 1 mg/g) and higher than those extracted by the non-sustainable methodology (15 ± 2 mg/g)	2020	[23]
Peach leaves	Ultrasound-assisted extraction (UAE) and Microwave-assisted extraction (MAE)	citric acid, xylitol and water (% m m^{-1}): 42:13:45) malic acid, xylitol and water (% m m^{-1}): 34:21:45) citric acid, malic acid and water (% m m^{-1}): 42:13:45)	UAE Extraction time—45 min Solid: liquid—90 mg of the samples: 9 mL of NADES Stirring rate—4000 rpm MAE Extraction temp - 100 °C Extraction time – 18 min Solid: liquid - 90 mg of the samples and 9 mL of NADES	The Analytical Eco-Scale scores - 97 per sample for UAE, and 96 per sample for MAE If the score > 75 represents an excellent green analysis	2020	[24]

(continued)

Table 1 (continued)

Type of plant materials	Type of extraction/ purification technique (s)	Composition and molar ratio of DES	Conditions of extraction	Summary of findings	Year	Reference
Soy protein	Liquid extraction	Choline chloride (HBA): Glycerol or ethylene glycol or 1,4-butanediol or 1,2-butanediol or polyethylene glycol (HBDs) molar ratio: 1:1 and 1:3	Extraction temp—60 °C Extraction time—3.9 h Solid: liquid—3:10 Stirring rate—873 rpm	Hydrogen bonding interactions within the DES influence protein extraction, while hydrophobic interactions also contribute to this process The DES induces protein denaturation and other alterations in the protein structure The extracted soy protein exhibited superior heat resistance compared to commercial protein alternatives	2021	[26]

(continued)

Table 1 (continued)

Type of plant materials	Type of extraction/ purification technique (s)	Composition and molar ratio of DES	Conditions of extraction	Summary of findings	Year	Reference
Oat protein	Liquid extraction	Choline chloride (HBA): Butanodiol isomers (HBDs) Molar ratio: 1:1, 1:2, and 1:3 Note: ChCl–1,4-butanediol-DES/ water binary mixture (molar ratio 1:3:1) was selected as novel extraction solvent	Extraction temp - 80 °C Extraction time – 90 min Solid: liquid - 1:9	Highest protein extraction - 55.28% DES mixture using molar ratio 1:3 formed liquid while others molar ratio formed precipitate High viscosity was formed after mixing DES with oat flour, limiting the efficiency of protein extraction The presence of water influenced the denaturation temperature (Td) and reflected the thermal stability of the protein against thermal processing	2021	[27]

(continued)

Table 1 (continued)

Type of plant materials	Type of extraction/ purification technique (s)	Composition and molar ratio of DES	Conditions of extraction	Summary of findings	Year	Reference
Oat protein	Liquid extraction	Choline chloride (HBA): 1,2-ethanediol, 1,3-propanediol, and 1,4-butanediol (HBDs) Molar ratio: 1:1, 1:2, and 1:3 Note: ChCl–1,3-propanediol-DES/ water binary mixture (molar ratio 1:2:1) was selected as solvent	Extraction temp - 80 °C Extraction time – 90 min Solid: liquid - 1:9	Highest protein extraction - 62.50% Longer chain lengths required higher concentrations to dissolve in HBA to form a clear liquid Increasing the chain length of dihydric alcohols in the DES showed favourable extraction profiles, with the addition of water did not affect the extraction performance	2022	[28]

(continued)

Table 1 (continued)

Type of plant materials	Type of extraction/ purification technique (s)	Composition and molar ratio of DES	Conditions of extraction	Summary of findings	Year	Reference
Pomegranate seeds protein	Ultrasound-assisted extraction Pressurized liquid extraction (PLE), under alkaline condition	Choline chloride (HBA): Ethylene glycol, glycerol, glucose, sorbitol, citric acid, sucrose, urea, or acetic acid (HBDs) Molar ratio: 1:1, 1:2, 2.1 or 4:1 Guanidine hydrochloride (HBA): Urea Molar ratio: 1:2	Extraction temp - 80 °C High intensity focused ultrasounds (HIFU) probe was utilized at 30% amplitude (39 W) for 5 min followed by centrifugation for 10 min at 4000 × *g* Solid: liquid – 6:1 Successive extraction using PLE: sodium bicarbonate buffer (pH 11.0), 170 °C for 3 cycles of 12 min each	Highest protein extraction - 61% Higher water content influenced the formation of strong hydrogen bonds and reduced the solubility of the extracted protein	2022	[29]
Sesame meal	Solid-Liquid extraction	Choline chloride (HBA): ethylene glycol (HBD) Molar ratio: 1:1	Extraction temp - 60 °C Extraction time – 60 min Solid: liquid - 1:20	Proteins extracted by DES are significantly better than proteins extracted by alkali-soluble acid	2023	[33]
Sacha inchi seed meal	Solid-Liquid extraction	Choline chloride (HBA): glycerol (HBD) Molar ratio: 1:2	Extraction temperature - 60 °C Extraction time – 60 min Solid: liquid - 1:20	The sacha inchi seed protein, extracted using ultrasound-ChCl/glycerol, showed remarkable solubility at alkaline pH and displayed the higher in vitro digestibility	2023	[30]

(continued)

Table 1 (continued)

Type of plant materials	Type of extraction/ purification technique (s)	Composition and molar ratio of DES	Conditions of extraction	Summary of findings	Year	Reference
Wheat germ	Solid-Liquid extraction	Potassium carbonate (HBA): anhydrous glycerol (HBD) Molar ratio: 1:3	Extraction temperature - 20 °C Extraction time – 3 min Solid: liquid - 1:30	The application of a DES buffer through electromagnetic-based microwave technology revealed a synergistic correlation between the absorbed microwave energy and the recoverable protein amount from defatted wheat germ	2023	[34]
Microalgae	Solid-Liquid extraction	Choline chloride (HBA): urea (HBD) Molar ratio: 1:2	Extraction temperature - NA Extraction time – NA Liquid: liquid - 8:1	The characterization indicates that both the target protein and the DES solution are effectively shielded within the DES-PC composite top phase. This protection enhances the feasibility of sustainability recycling the DES	2023	[35]

(continued)

Table 1 (continued)

Type of plant materials	Type of extraction/ purification technique (s)	Composition and molar ratio of DES	Conditions of extraction	Summary of findings	Year	Reference
Fava bean proteins	Solid-Liquid extraction	Choline chloride (HBA): glycerol (HBD) Molar ratio: 1:2	Extraction temperature - 50 °C Extraction time – 60 min Solid: liquid - 8:1	The extraction using DES led to an increase in the α-helix secondary structure component compared to proteins extracted through alkaline methods. However, the effect of DES extraction on protein denaturation was lower, implying potential superior functional properties in food systems	2024	[36]
Brewery spent grain	Solid-Liquid extraction	Tartarik acid (HBD): Choline chloride (HBA) Molar ratio: 1:2	Extraction temperature - 180 °C Extraction time – 20 min Solid: liquid - 1:10	Analysing a broad spectrum of bioactive compounds in Brewery spent grain through DES-mediated subcritical water hydrolysis can be advantageous for the effective utilization of BSG in various industrial applications	2024	[37]

powder was introduced into a two-phase DES system. The efficacy of the developed DES systems in isolating analytes was affirmed. Notably, choline chloride and tetra-butylammonium chloride in DES demonstrated robust phase-forming capabilities, generating weakly aqueous solutions. This property facilitated the stable transfer of compounds with varying polarities to the hydrophobic phase. The extraction process involved stirring at 150 rpm at a temperature of 25 °C for 10 min. Using a 10 mL glass syringe, the DES solvent was subjected to multiple push–pull cycles, resulting in the formation of turbid DES droplets. After centrifugation for 5 min, samples were withdrawn from the hydrophilic phase for content determination and from the hydrophobic phase for content evaluation. This methodology ensured efficient extraction of monosaccharides and amino acids from kelp, laying the foundation for the subsequent analytical assessments. From this study, the optimal concentrations of D-(+)-galactose, L-(−)-fucose, DL-tyrosine, and DL-valine in kelp were determined as 16.7 ± 0.2, 8.6 ± 0.2, 2.6 ± 0.1, and 1.6 ± 0.1 mg/g, respectively. Recovery rates for D-(+)-galactose, L-(−)-fucose, DL-tyrosine, and DL-tyrosine ranged from 87 to 102%, 84 to 103%, 87 to 104%, and 85 to 103%, respectively.

Meng et al. [22] conducted the extraction process by cleaning the kelp, followed by drying it at 50 °C and grinding it into a powder. Subsequently, 200 mg of the kelp powder was introduced into a two-phase DES system. The developed DES systems were demonstrated to be effective in isolating analytes. Specifically, choline chloride and tetra-butylammonium chloride in DES exhibited strong phase-forming abilities, resulting in the formation of weakly aqueous solutions that promoted the stable transfer of compounds with different polarities to the hydrophobic phase. Briefly, for the extraction, the mixture was stirred at 150 rpm at 25 °C for 10 min. By utilizing a 10 mL glass syringe, the DES was subjected to the solvent by multiple push–pull cycles, leading to the formation of turbid DES droplets. After centrifugation for 5 min, the samples were withdrawn from the hydrophilic phase for content determination and from the hydrophobic phase for content evaluation. This extraction methodology ensured the efficient retrieval of monosaccharides and amino acids from kelp, providing a robust foundation for subsequent analytical assessments. This extraction method demonstrated that the potential of the studied systems to offer an environmentally friendly extraction method for biomolecules in the fields of separation and purification.

Hernández-Corroto et al. [23] developed two green methods for the extraction of proteins and bioactive peptides from pomegranate peel based on pressurized liquid extraction (PLE) and a DES. The PLE extract and its hydrolysates exhibited notable antioxidant and hypocholesterolemic capacities, attributed to the coextraction of phenolic compounds and intact proteins. Conversely, hydrolysates from proteins extracted by DES displayed a high antihypertensive capacity, attributed to released peptides. In this study, eight different DES for protein extraction from pomegranate peel were prepared, consisting of a combination of hydrogen bond acceptor (HBA) and hydrogen bond donor (HBD) components. To control solvent viscosity, all DES formulations included water as the third component. The DES components were mixed at a 1:1:10 M ratio and heated in a water bath at 80 °C with agitation until a clear liquid was obtained. After preparation, the DES were used for protein extraction from pomegranate peel. SDS-PAGE analysis revealed distinctive band patterns, and based on these results, ChCl:AA DES was selected for further optimization using a Box-Behnken experimental design. High-Intensity Focused Ultrasound (HIFU) conditions were systematically adjusted, resulting in optimal parameters (ChCl:AA:H_2O DES at a 1:1:10 M ratio, HIFU amplitude 60%, 11 min) that yielded a protein extraction of 20 $\pm$ 1 mg/g peel. This yield surpasses that of the PLE method, highlighting the efficiency and potential industrial relevance of the developed green methodology.

Santana and co-workers [24] investigated the benefits of solvents with both low density and viscosity, emphasizing their advantage in achieving higher extraction efficiency through lower diffusivity and subsequently enhanced mass transfer. In this study, a mixture design was conducted to synthesize three NADES using citric acid, malic acid and xylitol. The optimized proportions for each of the three synthesized solvents were determined based on density and viscosity values. The NADES were subjected to characterization through infrared spectroscopy analysis, which revealed characteristic bands of the initial reagents and confirmed the presence of hydrogen bonds, indicating the formation of each DES. Subsequently, these NADES were employed as solvents in ultrasound-assisted extraction (UAE) and microwave-assisted extraction (MAE) processes for the determination of As, Cd, Hg, Pb, Se, and V using inductively coupled plasma mass spectrometry (ICP-MS). Santana and co-workers compared the results of their proposed methods to microwave-assisted acid digestion (MW-AD). The extraction recoveries for most of the determined elements ranged from 80 to 120%. Additionally, the researchers assessed the greenness of the proposed analytical procedures (UAE and MAE) using the Analytical Eco-Scale, indicating that these methods provided excellent green analyses, while the MW-AD method was rated as an acceptable green procedure.

Naseem et al. [25] developed a DES by combining choline chloride and glycerol, serving as an environmentally friendly alternative to conventional organic solvents for extraction purposes. The DES, along with aqueous methanol, was employed for ultrasound-assisted extraction (UAE) of bioactive compounds from *Mentha arvensis*, a plant known for its significant medicinal properties. The optimization of extraction parameters for phytochemical evaluation was conducted using response surface methodology. The extraction yield achieved with DES surpassed that obtained with aqueous methanol. Specifically, DES resulted in notable concentrations of total phenolic contents (TPCs) at 117 mg GAE/g, total flavonoids contents (TFCs) at 80 mg QE/g, and a remarkable 94% inhibition of 2,2-diphenyl-1-picrylhydrazyl (DPPH) radicals. The antimicrobial activities of the extracts were assessed against bacterial strains (S. aureus, E. coli) and fungal strains (F. solani, A. niger) using the well diffusion method. Characterization of the extracts through LC–MS revealed that chrysin, p-coumaric acid, naringenin, scopoletin, phenylpyruvic acid, pinocembrin, hesperidin, carnosic acid and caffeic acid were identified as the main bioactive components of *Mentha arvensis*. The DES exhibited a significantly high extraction yield, correlating with elevated TPCs, TFCs, and DPPH assay results compared to aqueous methanol. The extensive hydrogen bonding within the DES underscores its environmental significance in comparison to traditional solvents.

Chen et al. [26] examined the impact of polyol-based DES, prepared by mixing choline chloride (ChCl) (HBA) with five types of HBDs, namely glycerol, ethylene glycol, 1,4-butanediol, 1,2-butanediol, and polyethylene glycol, at specific HBA/HBD ratios, on the extraction of protein from soy. The protein was extracted using mechanical stirrer followed by centrifugation in order to collect supernatant containing protein extraction. Subsequently, the recovery step was proceeded with mixing supernatant containing protein with deionized water forming precipitate. At this stage the DES was denatured and cannot be recovered. It was observed that different combination of ChCl and HBDs resulted in different physicochemical characteristics and subsequently affect the interaction between the protein-based soy. HBD with large number of hydroxyl (OH) groups such as glycerol demonstrates excellent extraction efficiency compared to the other DESs due to enormous hydrogen bonding interactions with protein structures. Even though there are possibility for hydrophobic interactions to occur especially in HBDs containing longer chain lengths such as,4-butanediol, 1,2-butanediol, and polyethylene glycol, it was noticed that the stability during the interaction cannot be established effectively as a result of steric hindrance occurring between the DES and protein structure favouring more on hydrogen bonding. Through single-factor optimization experiments involving three variables (stirring speed, time, and liquid/solid ratio), optimal conditions were determined: extraction temperature of 60 °C, liquid/solid ratio of 10.3, stirring speed of 873 rpm, and time of 3.9 h. Furthermore, the extracted soy protein exhibited superior heat resistance compared to commercial protein alternatives.

Yue and co-workers [27] design an eco-friendly protein extraction method from oats by employing eighteen (18) formulations based on various configurations of ChCl-butanodiol-based DES isomers. In this study, all DESs prepared used different molar ratios (1:1, 1:2, and 1:3), either with or without the presence of water. During protein extraction, a mixture of oats and DES was incubated in a heating block at different times and temperatures of 80 °C, with intermittent shaking. The resulting supernatant containing extracted protein was centrifuged after cooling to room temperature and treated three times with DES solution before being centrifuged again. The collected oat protein precipitated upon addition of water. High viscosity was observed after mixing DES with oat flour, limiting the efficiency of protein extraction. Additionally, less polar mediums demonstrated good extraction performance, particularly observed with DESs containing HBDs such as 1,2-butanediol and 2,3-butanediol, compared to 1,4-butanediol. Extended extraction times beyond 90 min resulted in protein extraction being compromised due to the formation of beta-glucan, which increased the viscosity of the solution. The highest protein extraction, at 55.28%, was achieved using ChCl–1,4-butanediol-DES/water binary mixture (molar ratio 1:3:1). It was noted that the presence of water influenced the denaturation temperature (Td) and reflected the thermal stability of the protein against thermal processing. In another work by Yue et al. [28], they examines the influence of different chain lengths of dihydric alcohols (specifically 1,2-ethanediol, 1,3-propanediol, and 1,4-butanediol) on the structure of DES and its impact on protein extraction. The DESs were prepared by mixing HBA and three different HBDs at molar ratios of 1:1, 1:2, or 1:3, and heated at 100 °C with intermittent hand-shaking at 20-min intervals until a homogeneous and clear liquid was formed (the overall process takes about 60 min). For protein extraction, a mixture of oats and DES was dissolved at 80 °C for 90 min with a solid-to-liquid ratio of 1:9. It was noted that dihydric alcohols with longer chain lengths required higher concentrations to dissolve in HBA to form a clear liquid, facilitating strong and stable hydrogen bonding with HBA. Increasing the chain length of dihydric alcohols in the DES showed favourable extraction profiles, with the addition of water did not affect the extraction performance. Conversely, protein extraction lost its efficacy in anhydrous mixtures, with a reverse trend observed in ChCl/1,2-ethanediol-based DES, where optimal protein extraction occurred in an anhydrous mixture. The highest protein extraction, at 62.50%, was achieved with the ChCl–1,3-propanediol-DES/water binary mixture (molar ratio 1:2:1). Under optimal pH of 9, DES exhibited better functionality in protein extraction, governed by the ζ–potential.

Guzman-Lorite et al. [29] synthesized nine types of NADES consisting of various combinations of HBAs and HBDs (Fig. 1). Among these combinations, the DES formed by mixing choline chloride with acetic acid under constant stirring at 70–80 °C exhibited superior extraction performance compared to other prepared DESs. Utilizing High intensity focused ultrasounds (HIFU) or pressurized liquid extraction (PLE) with an alkaline buffer achieved an extraction yield of nearly 61% of protein from pomegranate seeds. The study noted that higher water content influenced the formation of strong hydrogen bonds and reduced the solubility of the extracted protein. Furthermore, protein extraction efficiency was compromised when using DESs formed with sugar-based HBDs like glucose, sucrose, or sorbitol, as well as glycerol-based DESs, due to oxidation and/or caramelization of NADES during the extraction process. Using Box-Behnken experimental design, optimal conditions were predicted for the extraction technique of proteins from pomegranate seeds: ultrasound amplitude of 80%, extraction time of 15 min, temperature of 60 °C, and sample amount of 27.8 mg.

DES possess a significant dissolution capability, rendering DES ideal for extracting various analytes, including macromolecules. Despite that, the choice of

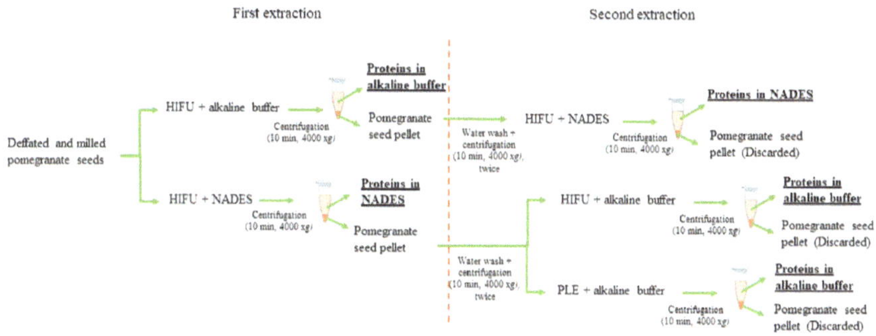

Fig. 1 Flow chart of the successive extraction of proteins using HIFU and PLE combined with alkaline buffer or NADES. Reprinted from [30], with permission from Elsevier

solvent can be tuned to enhance extraction performance, the selection of extraction method is crucial. The direct solid–liquid extraction approach is most commonly employed using DES [31]. The solvent choice plays a pivotal role in protein extraction. Both organic and aqueous solvents, as well as conventional and innovative techniques, have been utilized for extraction purposes over the years. However, challenges such as protein denaturation, alteration, environmental pollution, and toxicity have prompted further investigation. Recently, the use of DES for protein extraction has opened up new research avenues due to their sustainability, low economic cost, biodegradability, and nontoxicity [12]. This solvent can effectively address these challenges and yield consumable, natural proteins without any modifications [32].

Cao et al. [33] developed four DES systems (choline chloride: glycerol, choline chloride: ethylene glycol, choline chloride: oxalic acid, and choline chloride: glycolic acid) for the extraction of sesame protein. The developed DES was compared to the conventional method of alkali-soluble acid precipitation in terms of structure and function. The extracted protein with a purity of 93.3% was obtained using DES (choline chloride: ethylene glycol). Additionally, the choline chloride: oxalic acid system exhibited good solubility, emulsification, foaming, and thermal stability. Overall, these results indicate that DES can effectively extract sesame protein with characteristics suitable for use in the food industry. Moreover, the research lays the groundwork for the high-value utilization of sesame meal, currently discarded as waste. Sharma et al. [30] developed a novel protein extraction method based on ultrasound-assisted DES from sacha inchi seed meal. In this study, four different types of DES were prepared and investigated for extracting protein from sacha inchi meal. Choline chloride/glycerol (1:2) showed promising results compared to other three studied DES. The sequential ultrasound-ChCl:glycerol method could effectively extract a high total crude protein content (77.43%) from sacha inchi meal biomass, surprising the results of ultrasound (29.21%) or Chcl:glycerol (58.32%) treatment strategies. The method demonstrated significant potential for one-step biorefining of sacha inchi meal biomass for a circular bioeconomy. Olalere and Gan developed DES for protein extraction from defatted cereals, aiming to overcome the limitations of organic or ionic solvents [34]. Microwave-assisted protein extraction from defatted wheat germ was optimized using potassium carbonate:anhydrous glycerol (1:3). Under optimized condition, the protein yield increased while reducing microwave energy absorbance without any degradation of protein extracts. The maximum protein content and absorbable microwave energy were obtained as 33.00% and 677 J/min, respectively. The study revealed a synergistic correlation between the absorbed microwave energy and the amount of recovered protein from defatted wheat germ using the developed DES buffer via electromagnetic-based microwave technology. Zhuang et al. [35] developed the extraction of phycocyanin using DES-based liquid biphasic system. Phycocyanin is an intracellular protein produced in *Spirulina sp.* and is widely used as a natural dye in the cosmetic, food, and supplements industries. The highest phycocyanin extraction efficiency and yield were achieved when using choline chloride: urea, reaching an extraction efficiency of 94.21% and a yield of 92.03%.

Hewage et al. [36] developed a method to extract fava bean protein using an eco-friendly DES containing choline chloride and a glycerol-based system (1:2 w/w). This developed method was compared to the conventional approach using an alkaline solution to for fava bean protein extraction. Method conditions were optimized using the response surface methodology model: a liquid-to-solid ratio of 28:1 w/w, a choline chloride: glycerol molar ratio of 1:2 w/w, a temperature of 50 °C, an extraction time of 60 min, and a 40% w/w water content. The optimum extraction conditions resulted in protein content, protein yield, and protein recovery rate of 92.33%, 65.42%, and 23.25%, respectively. In contrast, alkaline extraction exhibited similar protein content (92.5%) but lower protein yield (60.76%) and recovery rates (21.74%) compared to DES-extracted proteins ($P < 0.05$). The DES formulation demonstrated potential as an extractor for fava bean protein, suitable for new food formulations. extraction of fava beans for use in new food formulations. Roy et al. [37] investigated the extraction performance of brewery spent grain (BSG) using subcritical water hydrolysis (SWH) modified by six different natural DES to determine their bipotentialities. The structural formula of HBA and HBD, as prepared, is shown in Fig. 2. This study showed that BSG hydrolysates obtained by SWH, with addition of safe catalytic agent DES, could be a promising method for obtaining high-value bioactive compounds. Industrial waste BSG was found to be an efficient source of protein, sugars, flavonoids, and phenolic compounds. DES-induced SWH BSG hydrolysates exhibited a protein content of 228 mg bovine serum albumin from per gram dried BSG. The extraction efficiency of bioactive of compounds depends on the type of DES, with DES-2 (tartaric acid: choline chloride in a 1:2 ratio) has shown better extraction efficiency for protein molecules in this research during the SWH.

The solid–liquid ratio plays a crucial role influencing the viscosity of DES and mass transfer. The hydrogen-bond interaction between DES and extracts holds significant importance. Consequently, determining the optimal solid/liquid ratio is vital to achieve lower DES viscosity, relatively weak hydrogen-bond interaction, and minimize the waste of extractants [38]. Another critical parameter in the extraction of plant wastes is the extraction time. Improved extraction yields are observed with increased extraction time, reaching a plateau after a certain duration. Additionally, the efficiency of DES extraction is influenced by ultrasonic power and microwave parameters when coupled with ultrasound-assisted extraction or microwave-assisted extraction. The outcomes of protein extraction depend on the properties of target compounds, necessitating adjustments to these parameters.

3 Challenges in DES-Assisted Extraction of Proteins

The extraction efficiency is an important variable for evaluating the performance of extraction methods. Based on the reported literature, researchers have prepared a variety of DES, optimized the DES composition, and extraction conditions to achieve more than 90% extraction efficiency. Among the variety of DES, they have prepared,

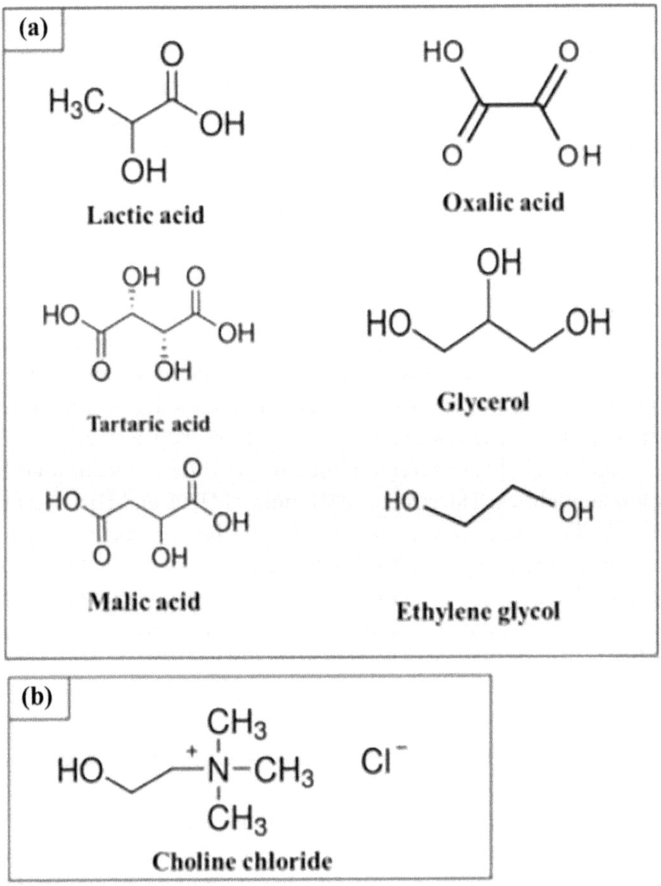

Fig. 2 The structural formulas of hydrogen bond donors (**a**) and the hydrogen bond acceptor (**b**) used to prepare natural deep eutectic solvents. Reprinted from [28], with permission from Elsevier

only a small portion has good extraction efficiency. This is due to the challenges in selecting DES components and conditions. One of the significant factors that contributed to the poor extraction efficiency is the high viscosity of DES due to the complex hydrogen bond network. This high viscosity could significantly affect the mass transfer process, heat transfer rate, and conductivity of the solvent, thereby hindering the flow of dissolved species. To overcome these shortcomings, several strategies can be employed to reduce the viscosity of DES, including adjusting the water content, temperature, and the molar ratio of components, while taking into account the influence of isomers on viscosity [18]. For example, the addition of water may reduce the viscosity of the DES, but excessive water can lead to the breakage of the hydrogen bond network in the system [39]. Most of the operational procedures to reduce viscosity are relatively tedious and complicated. Thus, another strategy to improve extraction efficiency is through the development of novel DES with low viscosity. The development of novel DES requires understanding the mechanism of structure in properties. For example, choline chloride-based DES and its analogues can be categorized into ionic type, and their viscosity depends on void volume. Therefore, the cavity theory is primarily applied to design for the improved DES [18]. The size of the cation should be small and the formed liquid should have a large void volume to obtain the ionic DES with acceptable and satisfactory viscosity [40]. In addition, hydrophobic DES (HDES) has been developed to overcome the limitation of hydrophilic DES and expanding the usefulness to extraction of non-polar compounds.

In reviewing the existing literature on DES for protein extraction from plants, the reported research mainly focuses on the extraction efficiency with limited investigation on extraction selectivity. Based on the principle of DES extraction, hydrophobic interaction, hydrogen bonding interaction, and salting out effect are the crucial factors that determine the transfer process. The extraction of specific protein could be accomplished by changing the important variables such as similar temperature and solid–liquid ratio. Hence, it is important to explore the mechanism of the impacts of different conditions on the extraction of protein using DES. Thus, certain DES components and extraction conditions are required for specific proteins can be employed to achieve specific extraction of target proteins [18, 41].

The issue of low back-extraction efficiency is also important to note, in addition to low extraction efficiency and specificity. It is known that protein can be back-extracted by changing the ionic strength, salt concentration and pH or adding ethanol to precipitate protein. However, these methods have not been widely employed and led to very low back-extraction efficiency. The development of efficient and reliable back-extraction methods may be the future research direction in the future. In addition, the combination of ultrasonic and microwave technologies, employing high-intensity ultrasound with DES, could contribute to the rapid separation of proteins, achieving high extraction efficiency, maintaining protein activity, and improving back-extraction efficiency [18, 21].

4 Conclusions

DES have been identified as a class of greener alternative solvents that have garnered much attention in research due to their favourable physicochemical properties, biodegradability, low toxicity, easy preparation, and sustainability. In addition to their wide applications in various research areas and industries, they are recently employed as greener solvents for efficiently separating and recovering proteins from plant materials. However, challenges have emerged in protein extraction using DES. Some of the issues that need to be addressed include extraction efficiency using low-viscosity DES, specificity issues in protein extraction, poor back-extraction efficiency, and the combination with microwave or ultrasound-assisted techniques. Thus, the optimization of DES preparation is crucial when designing protein extraction processes from plant sources. Further research should focus on the structure-property relationship to understand the mechanism behind low-viscosity DES, with a notable emphasis on scalability for industrial applications. This area is expected to rapidly develop in the future.

References

1. Kumar, M., Tomar, M., Potkule, J., Verma, R., Punia, S., Mahapatra, A., Belwal, T., Dahuja, A., Joshi, S., Berwal, M.K., Satankar, V., Bhoite, A.G., Amarowicz, R., Kaur, C., Kennedy, J.F.: Advances in the plant protein extraction: mechanism and recommendations. Food Hydrocoll. **115**(January), 106595 (2021). https://doi.org/10.1016/j.foodhyd.2021.106595
2. Tahergorabi, R., Hosseini, S.V.: Proteins, peptides, and amino acids. Elsevier, pp 15–38 (2017)
3. Pojić, M., Mišan, A., Tiwari, B.: Eco-innovative technologies for extraction of proteins for human consumption from renewable protein sources of plant origin. Trends Food Sci. Technol. **75**(October 2017), 93–104 (2018). https://doi.org/10.1016/j.tifs.2018.03.010
4. Singh, H., Sharma, R., Gupta, A., Joshi, S., Dar, B.N., Singh, B., Sharma, S.: Characterization of jackfruit seed enriched pasta: product-functionality profile, secondary protein structures, bioactive composition and molecular morphology. Qual. Assur. Saf. Crop. Foods **15**(2), 11–19 (2023). https://doi.org/10.15586/qas.v15i2.1217
5. Navaf, M., Sunooj, K.V., Aaliya, B., Sudheesh, C., Akhila, P.P., Mir, S.A., Nemtanu, M.R., George, J., Lackner, M., Mousavi Khaneghah, A.: Contemporary insights into the extraction, functional properties, and therapeutic applications of plant proteins. J. Agric. Food Res. **14**(December) (2023). https://doi.org/10.1016/j.jafr.2023.100861
6. Liu, R.-L., Yu, P., Ge, X.-L., Bai, X.-F., Li, X.-Q., Fu, Q.: Establishment of an aqueous PEG 200-based deep eutectic solvent extraction and enrichment method for Pumpkin (Cucurbita moschata) seed protein. Food Anal. Methods **10**(6), 1669–1680 (2017). https://doi.org/10.1007/s12161-016-0732-y
7. Dash, P., Ananthanarayan, L.: Development of Kesari dal (Lathyrus sativus) protein hydrolysates, with reduced β-ODAP content exhibiting anti-oxidative and anti-diabetic properties. J. Food Meas. Charact. **14**(4), 2108–2125 (2020). https://doi.org/10.1007/s11694-020-00458-5
8. Tan, M., Nawaz, M.A., Buckow, R.: Functional and food application of plant proteins—a review. Food Rev. Intl. **39**(5), 2428–2456 (2023). https://doi.org/10.1080/87559129.2021.1955918
9. Islam, M.S., Wang, H., Admassu, H., Sulieman, A.A., Wei, F.A.: Health benefits of bioactive peptides produced from muscle proteins: antioxidant, anti-cancer, and anti-diabetic activities. Process Biochem. **116**(March), 116–125 (2022). https://doi.org/10.1016/j.procbio.2022.03.007
10. Coelho, M.S., Salas-Mellado, M., de Las M.: How extraction method affects the physicochemical and functional properties of chia proteins. Lwt **96**(May), 26–33 (2018). https://doi.org/10.1016/j.lwt.2018.05.010
11. Gençdağ, E., Görgüç, A., Yilmaz, F.M.: Recent advances in the recovery techniques of plant-based proteins from agro-industrial by-products. Food Rev. Intl. **37**(4), 447–468 (2021). https://doi.org/10.1080/87559129.2019.1709203
12. Patra, A., Arun Prasath, V., Pandiselvam, R.: Deep eutectic solvent: an emerging trend for extraction of plant proteins. J. Mol. Liq. **389**, 122887 (2023). https://doi.org/10.1016/j.molliq.2023.122887
13. Bose, U., Broadbent, J.A., Byrne, K., Hasan, S., Howitt, C.A., Colgrave, M.L.: Optimisation of protein extraction for in-depth profiling of the cereal grain proteome. J. Proteomics **197**(November 2018), 23–33 (2019). https://doi.org/10.1016/j.jprot.2019.02.009

14. Cui, Q., Ni, X., Zeng, L., Tu, Z., Li, J., Sun, K., Chen, X., Li, X.: Optimization of protein extraction and decoloration conditions for tea residues. Hortic. Plant J. **3**(4), 172–176 (2017). https://doi.org/10.1016/j.hpj.2017.06.003
15. Dai, Y., Rozema, E., Verpoorte, R., Choi, Y.H.: Application of natural deep eutectic solvents to the extraction of anthocyanins from Catharanthus roseus with high extractability and stability replacing conventional organic solvents. J. Chromatogr. A **1434**, 50–56 (2016). https://doi.org/10.1016/j.chroma.2016.01.037
16. Contreras, M., del M., Lama-Muñoz, A., Manuel Gutiérrez-Pérez, J., Espínola, F., Moya, M., Castro, E.: Protein extraction from agri-food residues for integration in biorefinery: Potential techniques and current status. Bioresour. Technol. **280**(December 2018), 459–477 (2019). https://doi.org/10.1016/j.biortech.2019.02.040
17. Lores, H., Romero, V., Costas, I., Bendicho, C., Lavilla, I.: Natural deep eutectic solvents in combination with ultrasonic energy as a green approach for solubilisation of proteins: application to gluten determination by immunoassay. Talanta **162**(October 2016), 453–459 (2017). https://doi.org/10.1016/j.talanta.2016.10.078
18. Zhou, Y., Wu, W., Zhang, N., Soladoye, O.P., Zhang, Y., Fu, Y.: Deep eutectic solvents as new media for green extraction of food proteins: opportunity and challenges. Food Res. Int. **161**(August) (2022). https://doi.org/10.1016/j.foodres.2022.111842
19. Shishov, A., Bulatov, A., Locatelli, M., Carradori, S., Andruch, V.: Application of deep eutectic solvents in analytical chemistry, a review. Microchem. J. **135**, 33–38 (2017). https://doi.org/10.1016/j.microc.2017.07.015
20. Liu, Y., Friesen, J.B., McAlpine, J.B., Lankin, D.C., Chen, S.N., Pauli, G.F.: Natural deep eutectic solvents: properties, applications, and perspectives. J. Nat. Prod. **81**(3), 679–690 (2018). https://doi.org/10.1021/acs.jnatprod.7b00945
21. Li, G., Row, K.H.: Air assisted dispersive liquid-liquid microextraction (AA-DLLME) using hydrophilic-hydrophobic deep eutectic solvents for the isolation of monosaccharides and amino acids from kelp. Anal. Lett. **53**(2), 188–202 (2020). https://doi.org/10.1080/00032719.2019.1643358
22. Meng, J., Wang, Y., Zhou, Y., Chen, J., Wei, X., Ni, R., Liu, Z., Xu, F.: Development of different deep eutectic solvent aqueous biphasic systems for the separation of proteins. RSC Adv. **9**(25), 14116–14125 (2019). https://doi.org/10.1039/c9ra00519f
23. Hernández-Corroto, E., Plaza, M., Marina, M.L., García, M.C.: Sustainable extraction of proteins and bioactive substances from pomegranate peel (Punica granatum L.) using pressurized liquids and deep eutectic solvents. Innov. Food Sci. Emerg. Technol. **60**(February) (2020). https://doi.org/10.1016/j.ifset.2020.102314
24. Santana, A.P.R., Andrade, D.F., Guimarães, T.G.S., Amaral, C.D.B., Oliveira, A., Gonzalez, M.H.:. Synthesis of natural deep eutectic solvents using a mixture design for extraction of animal and plant samples prior to ICP-MS analysis. Talanta **216**(April) (2020). https://doi.org/10.1016/j.talanta.2020.120956
25. Naseem, Z., Zahid, M., Hanif, M.A., Shahid, M.: Environmentally friendly extraction of bioactive compounds from mentha arvensis using deep eutectic solvent as green extraction media. Pol. J. Environ. Stud. **29**(5), 3749–3757 (2020). https://doi.org/10.15244/pjoes/114235
26. Chen, Q., Chaihu, L., Yao, X., Cao, X., Bi, W., Lin, J., Chen, D.D.Y.: Molecular property-tailored soy protein extraction process using a deep eutectic solvent. ACS Sustain. Chem. Eng. **9**(30), 10083–10092 (2021). https://doi.org/10.1021/acssuschemeng.1c01848
27. Yue, J., Zhu, Z., Yi, J., Lan, Y., Chen, B., Rao, J.: Structure and functionality of oat protein extracted by choline chloride–dihydric alcohol deep eutectic solvent and its water binary mixtures. Food Hydrocoll. **112**(September 2020) (2021). https://doi.org/10.1016/j.foodhyd.2020.106330
28. Yue, J., Zhu, Z., Yi, J., Li, H., Chen, B., Rao, J.: One-step extraction of oat protein by choline chloride-alcohol deep eutectic solvents: role of chain length of dihydric alcohol. Food Chem. **376**(July 2021) (2022). https://doi.org/10.1016/j.foodchem.2021.131943
29. Guzmán-Lorite, M., Marina, M.L., García, M.C.: Successive extraction using natural deep eutectic solvents and pressurized liquids for a greener and holistic recovery of proteins from

pomegranate seeds. Food Res. Int. **161**(May) (2022). https://doi.org/10.1016/j.foodres.2022.111862
30. Sharma, V., Tsai, M.-L., Sun, P.-P., Chen, C.-W., Nargotra, P., Dong, C.-D.: Sequential ultrasound assisted deep eutectic solvent-based protein extraction from Sacha inchi meal biomass: towards circular bioeconomy. J. Food Sci. Technol. **60**(4), 1425–1434 (2023). https://doi.org/10.1007/s13197-023-05689-0
31. Kaoui, S., Chebli, B., Zaidouni, S., Basaid, K., Mir, Y.: Deep eutectic solvents as sustainable extraction media for plants and food samples: a review. Sustain. Chem. Pharm. **31**(December 2022) (2023). https://doi.org/10.1016/j.scp.2022.100937
32. Al-Maari, M.A., Hizaddin, H.F., Hayyan, A., Hadj-Kali, M.K.: Screening deep eutectic solvents as green extractants for oil from plant seeds based on COSMO-RS model. J. Mol. Liq. **393**, 123520 (2024). https://doi.org/10.1016/j.molliq.2023.123520
33. Cao, P.H., Zhang, C.X., Ma, Y.X., Yu, Y.M., Liu, H.M., Wang, X.D., Zheng, Y.Z.: Extraction of protein from sesame meal: Impact of deep eutectic solvents on protein structure and functionality. Lwt **187**(100) (2023). https://doi.org/10.1016/j.lwt.2023.115366
34. Olalere, O.A., Gan, C.Y.: Process optimisation of defatted wheat germ protein extraction in a novel alkaline-based deep eutectic solvent (DES) via Box–Behnken experimental design (BBD). Food Chem. **409**(November 2022) (2023). https://doi.org/10.1016/j.foodchem.2022.135224
35. Zhuang, D., Chew, K.W., Teoh, W.Y., Al-Maari, M.A.S., Hizaddin, H.F., Alharthi, S., Show, P.L., Ling, T.C.: Extraction of phycocyanin from spirulina using deep eutectic solvent liquid biphasic system. J. Taiwan Inst. Chem. Eng. **151**, 105125 (2023). https://doi.org/10.1016/j.jtice.2023.105125
36. Hewage, A., Olatunde, O.O., Nimalaratne, C., House, J.D., Aluko, R.E., Bandara, N.: Improved protein extraction technology using deep eutectic solvent system for producing high purity fava bean protein isolates at mild conditions. Food Hydrocoll. **147**(July 2023) (2024). https://doi.org/10.1016/j.foodhyd.2023.109283
37. Roy, V.C., Park, J.S., Haque, A.R., Ali, M.S., Lee, H.J., Chun, B.S.: Bio-refinery of brewery spent grain utilizing natural deep eutectic solvent-induced subcritical water. J. Supercrit. Fluids **204**(November 2023) (2024). https://doi.org/10.1016/j.supflu.2023.106108
38. Tang, W., Wu, Y., Wang, M., Row, K.H., Qiu, H., Zhou, J.-L.: Emerging application of extraction phase of ionic and non-ionic deep eutectic solvents toward natural herbal medicine. TrAC Trends Anal. Chem. **165**, 117137 (2023). https://doi.org/10.1016/j.trac.2023.117137
39. Wang, M., Wang, J., Zhang, Y., Xia, Q., Bi, W., Yang, X., Chen, D.D.Y.: Fast environment-friendly ball mill-assisted deep eutectic solvent-based extraction of natural products. J. Chromatogr. A **1443**, 262–266 (2016). https://doi.org/10.1016/j.chroma.2016.03.061
40. Alkhatib, I.I.I., Bahamon, D., Llovell, F., Abu-Zahra, M.R.M., Vega, L.F.: Perspectives and guidelines on thermodynamic modelling of deep eutectic solvents. J. Mol. Liq. **298**, 112183 (2020). https://doi.org/10.1016/j.molliq.2019.112183
41. Lin, J., Xiang, H., Sun-Waterhouse, D., Cui, C., Wang, W.: Deep eutectic solvents and alkaline extraction of protein from seabuckthorn seed meal: a comparison study. Food Sci. Hum. Wellness **11**, 1028–1035 (2022)

Pretreatment of Lignocellulosic Biomass Using Deep Eutectic Solvents

Ahmad Anas Nagoor Gunny, Nor Helya Iman Kamaludin, Nurul Hani Noor Asmadi, and Zaidatul Nadiah Zulkiflee

Abstract An abundant and sustainable resource, lignocellulosic biomass (LB) offers great potential for the production of value-added chemicals and biofuel. However, a proper enzymatic hydrolysis is limited by the LB's intricate, and complex structure. Environmental and financial concerns are raised by the use of harmful substances, extreme temperatures, and large energy inputs in conventional pretreatment procedures. A sustainable substitute for LB pretreatment, deep eutectic solvents (DES), which were first proposed as green solvents, combine efficient lignin and hemicellulose breakdown with low toxicity and a less environmental impact. This chapter covers the properties, mechanisms, and effectiveness of DES in biomass breakdown, emphasising how it helps to get over LB's structure resistance. Along with the benefits of incorporating DES with other pretreatment methods like mechanical refining and microwave irradiation, important factors affecting DES pretreatment are explained, such as temperature, composition, biomass loading, and reaction time. DES is a promising option for sustainable biorefinery applications owing to its capacity to promote enzymatic hydrolysis and increase sugar yields. Considering its benefits, DES recycling techniques that are more cost-effective, scalable, and optimised are still essential for wider industrial use.

1 Introduction

Lignocellulosic biomass (LB) is one of the most abundant biopolymer and renewable energy resources for the production of biofuel and other bio-products [1]. The major components of lignocellulosic biomass are polymeric carbohydrates which consist of cellulose (30–50 wt %), hemicelluloses (15–35 wt %) and lignin (10–30

A. A. N. Gunny (✉) · N. H. I. Kamaludin · N. H. N. Asmadi · Z. N. Zulkiflee
Faculty of Chemical Engineering Technology, Universiti Malaysia Perlis (UniMAP), Arau Perlis, Malaysia
e-mail: ahmadanas@unimap.edu.my

A. A. N. Gunny · N. H. I. Kamaludin
Centre of Excellence for Biomass Utilization (CoEBU), Universiti Malaysia Perlis (UniMAP), Arau Perlis, Malaysia

D. N. Jimat (ed.), *Deep Eutectic Solvents*, Engineering Materials,
https://doi.org/10.1007/978-981-95-1912-5_7

wt %). Cellulose represents the largest part of lignocellulose which can be hydrolysed to generate monomeric sugars units. These sugars can be converted via the bio-processing approach to a wide range of value-added chemicals.

In general, the conversion of lignocellulosic biomass to desired products consists of three major steps: (1) Lignocellulose is pre-treated to bring the sugar polymers into a form suitable for hydrolysis, (2) The enzymatic hydrolysis of cellulose to fermentable sugar and (3) The fermentation of simple sugar into desired products [2]. However, the physicochemical and compositional features of lignocelluloses limit their enzymatic hydrolysis process. Structural complexity is a major barrier for green and cost effective technology for processing lignocellulosic materials to sugars and other bio-products [3]. Thus, the pre-treatment process of lignocelluloses is a crucial step in this conversion.

Typically, the pre-treatment of lignocelluloses is classified into physical, chemical, physico-chemical and biological pre-treatments. Different pre-treatment methods adopted different strategies and mechanism for the deconstruction of highly recalcitrant lignocellulosic biomass structures. However, conventional pre-treatment methods have some environmental issues and drawbacks. These include the use of hazardous chemicals and the emission of hazardous byproducts from conventional biomass pre-treatment methods. In addition, some of the methods need to be operated in severe conditions and have a high energy demand like high pressures, temperatures, acidic and alkaline conditions. This can make the pre-treatments environmentally detrimental and expensive [4]. Consequently, more efficient and environmentally benign pre-treatments methods are deemed necessary.

In recent decades, Deep Eutectic Solvent (DES) been emerged and acknowledged as the most promising green solvent in the biological and chemical applications. DES was first introduced by Abbott et al. [5]. DES is defined as a homogenous mixture composed of hydrogen bond donor (HBD) and hydrogen bond acceptor (HBA) that are able to self-associate by forming a new eutectic phase, with a melting point lower than its individual components [6]. DESs exhibit some common characteristics which are similar to those of the ionic liquids such as high solubility, low volatility, low melting point, low vapour pressure, dipolar nature, chemically and thermally stable. DES is easy to prepare by simply mixing the HBD and HBA at moderate temperature without post-synthesis purification [7]. All these properties allow the DES to be designed and fine-tuned in specific applications such as lignocellulose pretreatment for fermentable sugar production [8], bioethanol production [9] and biodiesel production [10].

The design and preparation of environmentally benign solvents in lignocellulose pretreatment still pose a great challenge for the researchers in order to achieve the fundamental principles of green chemistry [11]. In this case, the emergence of DES plays an important role in replacing the conventional solvents which are applied in lignocellulosic biomass hydrolysis owing to its excellent physicochemical properties. This enables DES to be modified and tailored based on the desired application. Therefore, it is worth mentioning that the DES might be the key factor for a greener and sustainable biochemical processing of lignocellulosic biomass in the biorefinery concept. In the following subsequent subtopics, the chapter will discuss structure of lignocellulosic biomass, current pretreatment method and its drawback and the advantages of pretreatment of biomass using DES and its unique properties.

2 Composition and Structure of Lignocellulosic Biomass

Lignocellulosic biomass (LB) is a rich and sustainable resource derived from plants, mostly consisting of an aromatic polymer (lignin) and polysaccharides (cellulose and hemicelluloses). Figure 1 shows the structure of LB.

Without risking the security of the world's food supply, LB offers a great deal of potential as a fossil resource substitute for the production of second-generation biofuels and bio sourced chemicals and minerals. The composition of LB is exceptionally based on the sources whether it is acquired from the wood or agricultural wastes such as onion skin [13], tomato peels [14], rice straw [15] and more. In general, agricultural LB usually made up of 10–25% of lignin, 20–30% of hemicellulose and 40–50% of cellulose [16]. Cellulose is a primary polysaccharide found in plant cell walls that made up of repeating (1,4)-D-glucopyranose units joined by β-1,4 links and having an average molecular weight of approximately 100,000 g/mol [17]. It is embedded in the matrix which makes it invulnerable toward enzymatic hydrolysis. Hemicellulose is an amorphous structure that is made up by various monosaccharide that act as a physical barrier to limit the accessibility of enzymes. Lignin is a highly

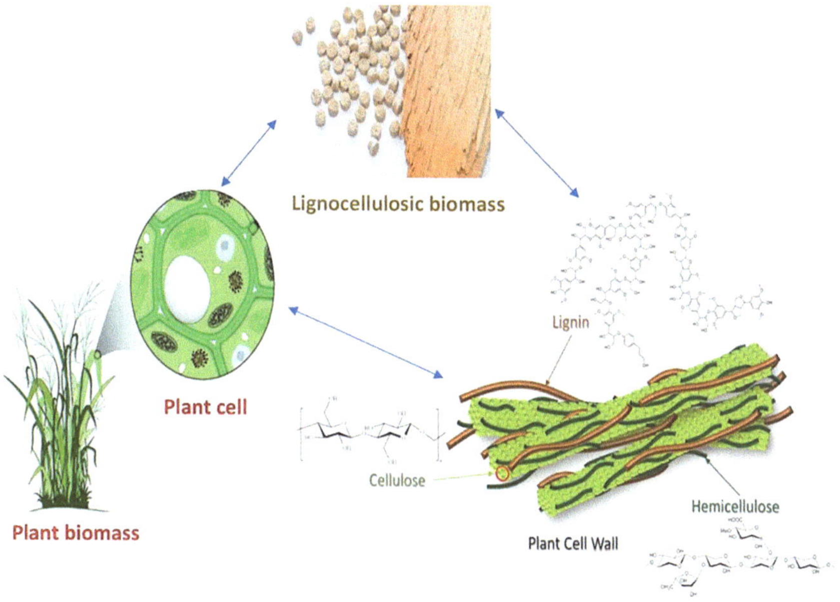

Fig. 1 Structure of lignocellulosic biomass [12]

intricate amorphous heteropolymer composed of phenylpropanoid building blocks (p-coumaryl, coniferyl, and sinapyl alcohol) [18]. The hydrophobicity and structural rigidity of the lignocellulosic matrix are attributed to lignin as it binds the hemicelluloses to cellulose in the cell wall. It is commonly recognized that lignin, depending on a number of criteria including its overall concentration and composition (particularly the amount of hydroxyl groups and S and G units), has a detrimental effect on the conversion of cellulose [19].

The aim of pretreatment is to convert biomass subject to enzyme hydrolysis of polysaccharides into fermentable sugars. However, the presence of lignin in LB can caused limitation in accessibility of polysaccharide as it acts as physical barriers that blocks the access of enzymes to cellulose. Due to its hydrophobic structural characteristics, which include hydrogen bonds, methoxy groups, and polyaromatic structures, the lignin can adsorb cellulases and other enzymes during enzymatic hydrolysis. Hence, an effective pretreatment is required to ensure a fully delignification of LB to allow polysaccharide accessibility during enzyme hydrolysis.

3 Overview of Conventional Biomass Pretreatment Methods

In the last few years, various studies have been conducted on the method of lignocellulosic biomass pretreatment such as alkaline pretreatment, bleaching, extrusion and more. During these studies, several agricultural wastes has been used such as rice straw [15], garlic straw [20], kenaf fibre [21] and more. The following Table 1 is an overview of the conventional Biomass Pretreatment.

4 Deep Eutectic Solvents: Characteristics and Properties

Deep Eutectic Solvents (DESs) have emerged as promising alternatives to traditional solvents, particularly in biological and chemical applications. Initially introduced by [5], DESs are defined as homogeneous mixtures of a hydrogen bond donor (HBD) and a hydrogen bond acceptor (HBA). They are notable for having a melting point that is significantly lower than that of their individual components [27, 28]. This melting point depression is due to the charge delocalization resulting from hydrogen bonding between the HBD and HBA [29]. The first well-known DES, "reline," combines choline chloride with urea, yielding a mixture with a melting point of approximately 12 °C, far below that of the individual components [5]. DESs are frequently compared to ionic liquids (ILs) because they share similar properties such as low volatility and high thermal stability. However, DESs offer additional benefits, including lower toxicity, enhanced biodegradability, and reduced cost [30]. Their ease of synthesis

Table 1 Conventional biomass pretreatment methods

Pretreatment methods	Pretreatment process	Lignin content before and after pretreatment (%)	Limitation	References
Alkaline pretreatment	Alkaline solution: 4% NaOH Pulping temperature: 80 °C Pulping time: 2 h	Before: 20.24 After: 15.73	– Highly corrosive chemical – Tend to cause alteration in lignin structure	[22]
Alkaline pretreatment	Alkaline solution: 2% NaOH Pulping temperature: Room temperature Pulping time: 12 h	Before: 6.3 After: 4.6		[20]
Alkaline pretreatment	Alkaline solution: 2% NaOH Pulping temperature: 100 °C Pulping time: 4 h	Before: 32.29 After: 18.73		[23]
Bleaching	Bleaching solution: 1.7% sodium chlorite Bleaching temperature: 70–80 °C Bleaching time: 2 h	Before: 18.79 After: 1.05	– Harsh and toxic chemical usage – High in cost – Disposal of bleaching chemical can have a significant environmental effect if does not properly treated	[24]
Bleaching	Bleaching solution: Aqueous chlorite and acetate buffer Bleaching temperature: 80 °C Bleaching time: 4 h	Before: 5.5 After: 1.8		[25]

(continued)

Table 1 (continued)

Pretreatment methods	Pretreatment process	Lignin content before and after pretreatment (%)	Limitation	References
Bleaching	Bleaching solution: 2% sodium chlorite -3% acetic acid mixture 1.5% NaOH-1% hydrogen peroxide mixture 1.25% sodium chlorite -3% acetic acid mixture Bleaching temperature: 70 °C Bleaching time: 6 h	Before: 12.7 After: 0.5		
Extrusion	Reagent: NaOH + sulfuric acid Temperature: 110 °C Screw speed: 100 rpm	Before: 1.54% After: 1.51	– Tend to consume high energy – Expensive equipment – Limited lignin removal	[26]

and the use of renewable materials make them attractive for applications such as lignocellulose pretreatment, organic synthesis, and biomass hydrolysis [31, 32].

DESs are characterized by several distinctive properties that contribute to their effectiveness as green solvents. A primary feature is their low melting point, achieved through the formation of strong hydrogen bonds between the HBD and HBA, making them suitable for various uses including as solvents and electrolytes [5]. Additionally, DESs exhibit high thermal stability, minimal vapor pressure, and low volatility, which are beneficial for processes requiring stable and high-temperature solvents [33]. Another significant property of DESs is their adjustable polarity. By varying the HBD and HBA components, the solvent's solubility and other properties can be tailored to dissolve a wide range of substances, enhancing their versatility in organic synthesis and material processing [34]. DESs are also noted for their straightforward and cost-effective preparation. Typically, DESs are created by heating and mixing the HBD and HBA under an inert atmosphere until a homogeneous liquid forms, avoiding the need for additional solvents or complex purification steps [35].

DESs are distinguished by several characteristics that make them particularly appealing as sustainable solvent alternatives. A key characteristic is their environmentally friendly nature, as DESs are often derived from natural and renewable resources like choline chloride and urea, which are both affordable and biodegradable [36, 37]. For example, choline chloride, a common HBA, is a vitamin B component produced in large quantities for animal nutrition [38], while urea is widely used in fertilizers [39]. Additionally, DESs are known for their low toxicity and ease of preparation. Unlike many ionic liquids, which can be expensive and complex to produce, DESs can be synthesized relatively simply by mixing the HBD and HBA components under mild conditions. This simplicity, combined with their renewable nature, makes DESs cost-effective and practical for large-scale industrial use [30]. DESs also possess ionic properties that allow them to dissolve ionic and polar compounds, which, together with their low volatility and high thermal stability, makes them suitable for a variety of applications including metal extraction and biocatalysis [40]. The ability to customize DESs by adjusting the HBD and HBA components adds to their versatility, making them effective for specific industrial and chemical processes [41].

In the field of biomass processing, DES are gaining attention for their potential to address the challenges associated with transforming raw plant materials into valuable chemical products. Biomass, including plant residues, represents a plentiful, affordable, and renewable resource. Traditionally, biomass has been primarily used for energy production through direct combustion. However, it also contains valuable components like cellulose, hemicellulose, and polysaccharides that can be converted into biofuels and biochemicals via fermentation [42]. A significant portion of biomass, primarily composed of lignin and biopolymers, remains underutilized (10–30% of the plant mass), indicating an opportunity for more efficient utilization [43, 44]. Conventional biomass pretreatment methods, including physical, chemical, and biological techniques, often face challenges such as high energy demands, toxicity, and environmental concerns. These issues significantly increase the overall operational cost of converting lignocellulose into fermentable sugars [45–47]. DESs present a promising alternative, showing effectiveness in dissolving and extracting lignin and other components from lignocellulosic biomass. Recent studies demonstrated that a potassium carbonate-glycerol DES could isolate up to 73.8% cellulose from rice straw, using an optimized pulping method [48]. These results suggest that DESs could significantly enhance biomass processing efficiency, with further optimization of DES formulations potentially offering even greater improvements.

5 Parameters Affecting DES-Based Pretreatment Processes and Mechanisms of Biomass Deconstruction Using DES

DES pretreatment attempts to convert the complex carbohydrates into cellulose, hemicellulose, and lignin, in order to facilitate the enzymatic hydrolysis of cellulose and hemicellulose. Acidic DES is frequently utilized in pretreatment processes due to its effectiveness in eliminating lignin and hemicellulose, significantly enhancing the enzymatic hydrolysis of lignocellulosic pretreatment. Acid DES may effectively cleave the ether bond in lignin. The reaction begins with a proton attack in an acidic solvent on the oxygen atom of the α-position hydroxyl (or ether bond) on the benzene ring, resulting in the formation of a conjugate acid. The α-ether bond subsequently cleaves, a water molecule (or alkyl alcohol) is eliminated, and a stable resonance structure of a positively charged carbon is produced. Consequently, an elimination reaction involving the positive charge on the Cα and the hydrogen on the Cβ produces a carbon–carbon double bond and recovers a hydrogen ion. The hydrolysis of the carbon–carbon double bond subsequently results in the cleavage of the carbon–oxygen link, yielding 1-hydroxy-3-(4-hydroxy-3-methoxyphenyl)-2-propanone. Figure 2 shows the potential cleavage mechanism of lignin ether bonds in acid-catalyzed DES pretreatment.

The DES pretreatment efficiency usually affected by the parameters such as the temperature, DES composition, biomass loading, and reaction time used during the pretreatment process. The temperature used during pretreatment plays a critical role in enhancing the DES ability to breakdown the lignocellulosic biomass as generally, the increase in temperature will accelerate the dissolution of lignin and hemicellulose, hence, allowing more accessibility of cellulose for enzymatic hydrolysis. However, excessively elevated temperatures can result in the breakdown of carbohydrates,

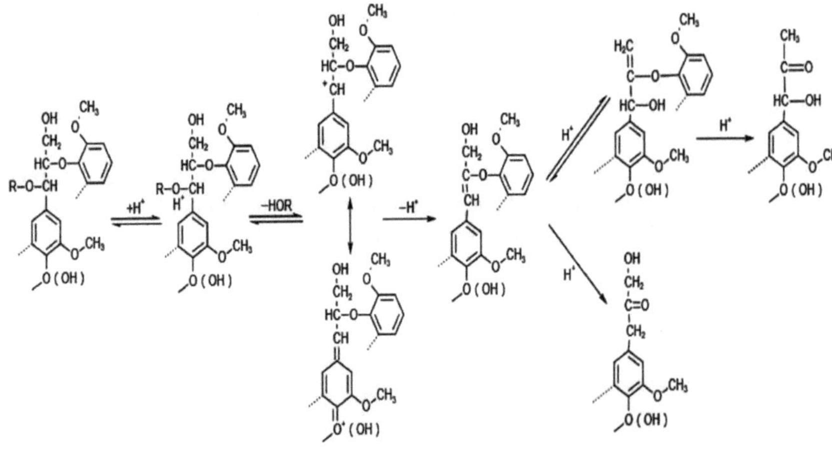

Fig. 2 Potential cleavage mechanism of lignin ether bonds in acid-catalyzed DES pretreatment [49]

including cellulose and hemicellulose, hence minimizing the quantity of cellulose. Therefore, an optimal temperature is needed to maximize the DES efficiency during pretreatment.

The DES composition also plays an important role as it can determine the solubility and reactivity of DES during pretreatment. Diverse hydrogen bond donors (HBDs) and acceptors (HBAs) possess different capacities to disrupt the lignin-carbohydrate matrix in biomass. For example, Sumiati et al. [50] reported that DES with HBD based on linear saturated acid such as formic acid and acetic acid has shown a significant dissolubility of lignin upon pretreatment compared to those containing neutral substances like urea. In addition, the ratio of HBA to HBD also affects the solvent's viscosity and solvation capacity, hence affecting the overall biomass deconstruction.

The biomass loading, defined as the quantity of biomass in relation to the volume of DES, directly influences the pretreatment process. Reduced biomass loadings generally result in enhanced lignin removal and increased cellulose exposure due to the solvent's better absorption into the biomass. Nonetheless, insufficient biomass loading leads to an increased amount of solvent utilized per unit of biomass, potentially rendering the process economically unsustainable.

The degree of biomass breakdown also affected by the duration of the pretreatment process where generally, an extended reaction periods typically lead to an enhanced elimination of lignin and hemicellulose, hence increasing cellulose accessibility. However, the prolonged exposure of DES on the biomass can result in the breakdown of cellulose and hemicellulose into inhibitory by-products, which may disrupt downstream processes such as fermentation. Therefore, an optimal reaction time for DES pretreatment is needed as short reaction times may result in insufficient pretreatment, whereas prolonged times can lead to excessive breakdown of the biomass which lead to degradation.

6 DES in Combination with Other Pretreatment Techniques

The pretreatment of lignocellulosic biomass is a crucial step in converting plant material into valuable products such as sugars, cellulose derivatives, and biofuels. Traditional pretreatment methods, such as enzymatic hydrolysis, mechanical refining, and size reduction, are often energy-intensive, expensive, and inefficient [51]. However, the introduction of DES has provided a promising solution for improving these pretreatment processes as demonstrated in past studies listed in Table 2. DESs are regarded as greener alternatives to conventional solvents, offering advantages such as lower toxicity, ease of preparation, and reduced environmental impact [52].

One of the key benefits of using DESs in combination with other pretreatment techniques is their ability to enhance enzymatic hydrolysis. Enzymatic hydrolysis is critical in breaking down cellulose into fermentable sugars, but the efficiency of this

Table 2 Pretreatment of lignocellulosic biomass using DES

DES	Pre-treatment process	Cellulose content	Additional treatment	Reference
I. Formic acid-choline chloride (1:2) II. Acetic acid-choline chloride (1:2) III. Lactic acid-choline chloride (1:2)	Temperature: 130 °C Time: 3 h Biomass to DES ratio: 1:20 (w/v)	Before: 43.2% After: I. Formic acid-choline chloride: 84.7% II. Acetic acid-choline chloride: 85.4% III. Lactic acid-choline chloride: 79.8%	Liquid hot water extraction	[53]
Potassium carbonate-glycerol (1:7)	Temperature: 140 °C Time: 1 h 40 minutes Biomass to DES ratio: 1 g:10 g	Before: 35.6% After: 73.8%	–	[48]
I. Formic acid-choline chloride (1:2) II. Sorbitol-choline chloride (1:2)	Temperature: 130 °C Time: 4 h Biomass to DES ratio: 1 g:10 g	Before: 40.13% After: I. Formic acid-choline chloride: 53.5% II. Sorbitol-choline chloride: 40.91%	–	[50]
Choline chloride-lactic acid (1:2)	Temperature:150 °C Time: 6 h Biomass to DES ratio: 1:20 (w/v)	Before: 36.77% After: 73.60%	–	[54]
I. Lactic acid-alanine (9:1) II. Formic acid-choline chloride (2:1) III. Glycolic acid-choline chloride (3:2)	Temperature: 60–150 °C Time: 10 min–24 h	Before: 40% After: I. Lactic acid-alanine: 57.8% II. Formic acid-choline chloride:52.8% III. Glycolic acid-choline chloride:49.6%	Ball milling and microwave irradiation	[55]

process is often limited by the recalcitrant nature of lignocellulosic biomass, especially due to lignin [56]. DESs have shown significant potential in removing lignin more effectively, thereby increasing cellulose accessibility to enzymes. Studies have demonstrated that DES pretreatment can improve enzymatic digestibility and significantly increase sugar yields compared to single-step pretreatment methods. For instance, [57] demonstrated that deep eutectic solvents (DES) in combination with commercial cellulases retained over 90% of enzyme activity after 1 h of incubation. This highlights the low inhibitory effects of DES on enzymes, reducing the need for extensive washing post-pretreatment. Additionally, the use of DES at higher pretreatment temperatures improved the production of reducing sugars, making the process more efficient. This synergy between DES and enzymatic hydrolysis effectively enhances sugar yields compared to single-step pretreatment methods.

Similarly, combining DES with mechanical refining processes has shown synergistic effects. Mechanical refining, which involves high energy input to break down biomass into smaller, more reactive particles, can be enhanced by DES pretreatment. The DES softens the biomass structure, reducing the energy needed for mechanical refining and increasing delignification [58, 59]. This combination results in improved sugar recovery during hydrolysis and makes the overall process more efficient. For example, a study by Ahorsu et al. [55] combining ball milling, microwave irradiation, and deep eutectic solvents (DES) has proven to be a highly effective method for rapidly removing lignin from materials with high lignin content, such as walnut shells. This process, which uses choline chloride and formic acid as the DES, can achieve high lignin removal rates of up to 80% in just 10 min at 150 °C. The process is more efficient than traditional methods, requiring less time and energy. The DES not only removes lignin but also preserves the polysaccharides with minimal degradation. This method shows promise for processing large volumes of tough biomass, making it a valuable addition to pretreatment techniques.

7 Conclusion

Due to the sustainability and versatility of DESs, they represent promising alternatives for future research, as well as industrial, applications. Due to their biodegradability, non-toxicity, and renewable origins, DESs provide a sustainable alternative to traditional hazardous solvents which can meet global demands for eco-friendly technological improvements across industries. DES systems can be designed directly with specific properties, providing an avenue for tailoring DES characteristics to meet desired attributes for a variety of industrial purposes from extraction, catalysis or enzyme stabilization. The fact that DESs are economic, and can be regenerated and reused with little loss of efficiency over multiple cycles makes them more attractive for industrial use on a large scale. The fact that they can work in less extreme conditions means big savings in energy, a crucial advantage for processes such as biomass pretreatment and chemical synthesis on bigger scales. Moreover, the selective dissolving of particular compounds by DESs makes them a nice candidate for extraction methods in the pharmaceuticals, cosmetic, and food industries, which require high purity and selectivity. In summary, it is quite apparent that DES has the potential to propel novelties in sustainable chemistry, biorefinery, and pharmaceutical scope. However, the current difficulties must be addressed with improved confidence and further research in order to fully utilize them for large-scale industrial applications.

The recent advancements in DESs for biomass pretreatment pave the way for greener and more efficient bioprocessing practices, surpassing conventional methods. This superiority is evident in their enhanced energy efficiency, achieved through milder operating conditions, and improved cellulose accessibility, driven by superior delignification and selectivity. Overall, the main trends follow the customization of DES compositions to improve selectivity for lignin, cellulose, and hemicellulose towards biomass fractionation with higher sugar yield. Research is also focused on combining DESs with energy-saving processes, such as microwave-assisted pretreatment or extrusion-assisted methods to improve the performance of lignocellulosic biomass pretreatment [60]. For example, microwave irradiation could positively influence ionic properties and molecular polarity of DES therefore improving fractionation yield as well as reducing reaction temperature and time without compromising process performance [61]. In contrast, DES-mediated extrusion effectively depolymerized and fibrillated biomass accompanied by the reduction in crystallinity and preservation of lignin structural features in carbohydrate [62]. Furthermore, one-pot DES in biorefinery has the potential to save energy notably since the solid/liquid separation step can be avoided during biomass saccharification [63].

In addition, more work is required to determine the potential DES recyclability and reusability at a higher number of deaerations in order for these systems to become more economically viable. In fact, potential strategies to improve the reusability and removal efficiency of recycled DESs in pretreatment processes are desirable for evaluating their performance. Moreover, a life cycle and techno-economic analysis of the entire process concerning to pretreatment using DESs are required to determine the overall feasibility. These assessments should consider the life-cycle costs of DES production, use, and recycling in relation to alternative pretreatment methods. Those studies are needed to determine if DES based processes are cost-efficient and sustainable over the long term, delivering critical information about their potential for industrial use.

Computational modeling is similarly advancing the mechanistic understanding of DES optimization. The application of molecular dynamics simulations with quantum chemical calculations, and the orientation of DESs to lignocellulosic biomass can be explored. This detailed knowledge can be useful to better tailor DES formulations around different types of biomass as well as enhance their performance and scalability. Collectively, these trends suggest that DESs will continue to develop as a sustainable and versatile platform for biomass pretreatment, with the potential to revolutionize the biofuel production and biorefinery processes. Nevertheless, scalability, solvent recovery, and process optimization are necessary for this technology to be implemented at full scale.

References

1. Pande, M., Bhaskarwar, A.N.: Biomass conversion to energy. In Baskar, C. (ed.), Biomass Conversion. Berlin Heidelberg Springer-Verlag (2012)
2. Liu, C.-Z., Wang, F., Stiles, A.R., Guo, C.: Ionic liquids for biofuel production: opportunities and challenges. Appl. Energy **92**, 406–414 (2012). https://doi.org/10.1016/j.apenergy.2011.11.031
3. Vancov, T., Alston, A.-S., Brown, T., McIntosh, S.: Use of ionic liquids in converting lignocellulosic material to biofuels. Renew. Energy **45**, 1–6 (2012). https://doi.org/10.1016/j.renene.2012.02.033
4. Mäki-Arvela, P., Anugwom, I., Virtanen, P., Sjöholm, R., Mikkola, J.P.: Dissolution of lignocellulosic materials and its constituents using ionic liquids—a review. Ind. Crops Prod. **32**(3), 175–201 (2010). https://doi.org/10.1016/j.indcrop.2010.04.005
5. Abbott, A.P., Capper, G., Davies, D. L., Rasheed, R.K., Tambyrajah, V.: Novel solvent properties of choline chloride urea mixtures. Chem. Commun. (0), 70–71 (2003). https://doi.org/10.1039/B210714G
6. Tomé, L.I.N., Baião, V., da Silva, W., Brett, C.M.A.: Deep eutectic solvents for the production and application of new materials. Appl. Mater. Today **10**, 30–50 (2018). https://doi.org/10.1016/j.apmt.2017.11.005
7. Francisco, M., van den Bruinhorst, A., Kroon, M.C.: Low-transition-temperature mixtures (LTTMs): a new generation of designer solvents. Angew. Chem.-Int. Ed. **52**(11), 3074–3085 (2013). https://doi.org/10.1002/anie.201207548
8. Procentese, A., Raganati, F., Olivieri, G., Russo, M.E., Rehmann, L., Marzocchella, A.: Deep eutectic solvents pretreatment of agro-industrial food waste. Biotechnol. Biofuels **11**(1), 37 (2018). https://doi.org/10.1186/s13068-018-1034-y
9. Xu, F., Sun, J., Wehrs, M., Kim, K.H., Rau, S.S., Chan, A.M., Simmons, B.A., Mukhopadhyay, A., Singh, S.: Biocompatible choline-based deep eutectic solvents enable one-pot production of cellulosic ethanol. ACS Sustain. Chem. Eng. **6**(7), 8914–8919 (2018). https://doi.org/10.1021/acssuschemeng.8b01271
10. Troter, D.Z., Todorović, Z.B., Đokić-Stojanović, D.R., Stamenković, O.S., Veljković, V.B.: Application of ionic liquids and deep eutectic solvents in biodiesel production: a review. Renew. Sustain. Energy Rev. **61**, 473–500 (2016). https://doi.org/10.1016/j.rser.2016.04.011
11. Gunny, A.A.N., Arbain, D., Javed, M., Baghaei-Yazdi, N., et al.: Deep eutectic solvents-halophilic cellulase system: an efficient route for *in situ* saccharification of lignocellulose. Process Biochem. (2019). https://doi.org/10.1016/j.procbio.2019.03.003
12. Swami, S., Suthar, S., Singh, R., Thakur, A.K., Gupta, L.R., Sikarwar, V.S.: Potential of ionic liquids as emerging green solvent for the pretreatment of lignocellulosic biomass. Environ. Sci. Pollut. Res. **31**(9), 12871–12891 (2024). https://doi.org/10.1007/s11356-024-32100-y
13. Rhim, J.W., Reddy, J.P., Luo, X.: Isolation of cellulose nanocrystals from onion skin and their utilization for the preparation of agar-based bio-nanocomposites films. Cellulose **22**(1), 407–420 (2015). https://doi.org/10.1007/s10570-014-0517-7
14. Jiang, F., Hsieh, Y.L.: Cellulose nanocrystal isolation from tomato peels and assembled nanofibers. Carbohyd. Polym. **122**, 60–68 (2015). https://doi.org/10.1016/j.carbpol.2014.12.064
15. Lu, P., Hsieh, Y.L.: Preparation and characterization of cellulose nanocrystals from rice straw. Carbohyd. Polym. **87**(1), 564–573 (2012). https://doi.org/10.1016/j.carbpol.2011.08.022
16. Shahzadi, T., Mehmood, S., Irshad, M., Anwar, Z., Afroz, A., Zeeshan, N., Rashid, U., Sughra, K.: Advances in lignocellulosic biotechnology: A brief review on lignocellulosic biomass and cellulases. Adv. Biosci. Biotechnol. **05**(03), 246–251 (2014). https://doi.org/10.4236/abb.2014.53031
17. Himmel, M.E., Ding, S.Y., Johnson, D.K., Adney, W.S., Nimlos, M.R., Brady, J.W., Foust, T.D.: Biomass recalcitrance: engineering plants and enzymes for biofuels production. Science **315**(5813), 804–807 (2007). https://doi.org/10.1126/science.1137016

18. Zoghlami, A., Paës, G.: Lignocellulosic biomass: understanding recalcitrance and predicting hydrolysis. Front. Chem. **7**(December) (2019). https://doi.org/10.3389/fchem.2019.00874
19. Santos, R.B., Lee, J.M., Jameel, H., Chang, H.M., Lucia, L.A.: Effects of hardwood structural and chemical characteristics on enzymatic hydrolysis for biofuel production. Biores. Technol. **110**, 232–238 (2012). https://doi.org/10.1016/j.biortech.2012.01.085
20. Kallel, F., Bettaieb, F., Khiari, R., García, A., Bras, J., Chaabouni, S.E.: Isolation and structural characterization of cellulose nanocrystals extracted from garlic straw residues. Ind. Crops Prod. **87**, 287–296 (2016). https://doi.org/10.1016/j.indcrop.2016.04.060
21. Jonoobi, M., Harun, J., Shakeri, A., Misra, M., Oksmand, K.: Chemical composition, crystallinity, and thermal degradation of bleached and unbleached kenaf bast (Hibiscus cannabinus) pulp and nanofibers. BioResources **4**(2), 626–639 (2009). https://doi.org/10.15376/biores.4.2.626-639.
22. Jiang, Y., Zhou, J., Zhang, Q., Zhao, G., Heng, L., Chen, D., Liu, D.: Preparation of cellulose nanocrystals from Humulus japonicus stem and the influence of high temperature pretreatment. Carbohyd. Polym. **164**, 284–293 (2017). https://doi.org/10.1016/j.carbpol.2017.02.021
23. Aridi, A.S., Chin, N., Ishak, N., Yusof, N., Kadota, K., Manaf, Y., Yusof, Y.: Effect of sodium hypochlorite concentration during pre-treatment on isolation of nanocrystalline cellulose from Leucaena leucocephala (Lam.) mature pods. BioResources **16**, 3137–3158 (2021). https://doi.org/10.15376/biores.16.2.3137-3158.
24. Susi, S., Ainuri, M., Wagiman, W., Falah, M.A.F.: Effect of delignification and bleaching stages on cellulose purity of oil palm empty fruit bunches. IOP Conf. Ser.: Earth Environ. Sci. **1116**(1) (2022). https://doi.org/10.1088/1755-1315/1116/1/012018
25. Rahman, N.H.A., Chieng, B.W., Ibrahim, N.A., Rahman, N.A.: Extraction and characterization of cellulose nanocrystals from tea leaf waste fibers. Polymers **9**(11), 1–12 (2017). https://doi.org/10.3390/polym9110588
26. Merci, A., Urbano, A., Grossmann, M.V.E., Tischer, C.A., Mali, S.: Properties of microcrystalline cellulose extracted from soybean hulls by reactive extrusion. Food Res. Int. **73**, 38–43 (2015). https://doi.org/10.1016/j.foodres.2015.03.020
27. Dai, Y., van Spronsen, J., Witkamp, G.-J., Verpoorte, R., Choi, Y.H.: Natural deep eutectic solvents as new potential media for green technology. Anal. Chim. Acta **766**, 61–68 (2013). https://doi.org/10.1016/j.aca.2012.12.019
28. Choi, Y.H., van Spronsen, J., Dai, Y., Verberne, M., Hollmann, F., Arends, I.W.C.E., Witkamp, G.-J., Verpoorte, R.: Are natural deep eutectic solvents the missing link in understanding cellular metabolism and physiology? Plant Physiol. **156**, 1701–1705 (2011). https://doi.org/10.1104/pp.111.178426
29. Abbott, A.P., Capper, G., Davies, D.L., Munro, H.L., Rasheed, R.K., Tambyrajah, V.: Preparation of novel, moisture-stable, Lewis-acidic ionic liquids containing quaternary ammonium salts with functional side chains. Chem Commun (2001). https://doi.org/10.1039/b106357j
30. Domínguez de María, P., Maugeri, Z.: Ionic liquids in biotransformations: from proof-of-concept to emerging deep eutectic solvents. Curr. Opin. Chem. Biol. **15**(2), 220–225 (2011). https://doi.org/10.1016/j.cbpa.2011.01.003
31. Gunny, A.A.N., Arbain, D., Daud, M.Z.M., Jamal, P.:. Synergistic action of deep eutectic solvents and cellulases for lignocellulosic biomass hydrolysis. Mater. Res. Innov. **18**(sup6), S6–67 (2014). https://doi.org/10.1179/1432891714z.000000000933
32. Francisco, M., Van Den Bruinhorst, A., Kroon, M.C.: New natural and renewable low transition temperature mixtures (LTTMs): screening as solvents for lignocellulosic biomass processing. Green Chem. **14**(8), 2153 (2012). https://doi.org/10.1039/c2gc35660k
33. Gorke, J.T., Srienc, F., Kazlauskas, R.J.: Hydrolase-catalyzed biotransformations in deep eutectic solvents. Chem. Commun. **44**, 1235–1237 (2008). https://doi.org/10.1039/B800282F
34. Abbott, A.P., Harris, R.C., Ryder, K.S., D'Agostino, C., Gladden, L.F., Mantle, M.D.: Glycerol eutectics as sustainable solvent systems. Green Chem. **13**(1), 82–90 (2011). https://doi.org/10.1039/C0GC00395F
35. Hansen, B.B., Spittle, S., Chen, B., Poe, D., Zhang, Y., Klein, J.M., Sangoro, J.R.: Deep eutectic solvents: a review of fundamentals and applications. Chem. Rev. (2020). https://doi.org/10.1021/acs.chemrev.0c00385

36. Khandelwal, S., Tailor, Y.K., Kumar, M.: Deep eutectic solvents (DESs) as eco-friendly and sustainable solvent/catalyst systems in organic transformations. J. Mol. Liq. **215**, 345–386 (2016). https://doi.org/10.1016/j.molliq.2015.12.012
37. Radošević, K., Cvjetko Bubalo, M., Gaurina Srček, V., Grgas, D., Landeka Dragicević, T., Radojčić Redovniković, I.: Evaluation of toxicity and biodegradability of choline chloride-based deep eutectic solvents. Ecotoxicol. Environ. Saf. **112**, 46–53 (2015). https://doi.org/10.1016/j.ecoenv.2014.09.034
38. Smith, E.L., Abbott, A.P., Ryder, K.S.: Deep eutectic solvents (DESs) and their applications. Chem. Rev. **114**(21), 11060–11082 (2014). https://doi.org/10.1021/cr300162p
39. Marsh, K.L., Sims, G.K., Mulvaney, R.L.: Availability of urea to autotrophic ammonia-oxidizing bacteria as related to the fate of 14C- and 15N-labeled urea added to soil. Biol. Fertil. Soils **42**(2), 137–145 (2005). https://doi.org/10.1007/s00374-005-0004-2
40. Pollet, P., Davey, E.A., Ureña-Benavides, E.E., Eckert, C.A., Liotta, C.L.: Solvents for sustainable chemical processes. Green Chem. **16**(2), 1034–1055 (2014). https://doi.org/10.1039/C3GC41440F
41. Garcia, G., Aparicio, S., Ullah, R., Atilhan, M.: Deep eutectic solvents: physicochemical properties and gas separation applications. Energy Fuels **29**(4), 2616–2644 (2015). https://doi.org/10.1021/acs.energyfuels.5b00093
42. Dutta, T., Papa, G., Wang, E., Sun, J., Isern, N.G., Cort, J.R., Simmons, B.A., Singh, S.: Characterization of lignin streams during ionic liquid-based pretreatment from grass, hardwood, and softwood. ACS Sustain. Chem. Eng. **6**(2), 3079–3090 (2018). https://doi.org/10.1021/acssuschemeng.7b04634
43. Kim, K.H., Eudes, A., Jeong, K., Yoo, C.G., Kim, C.S., Ragauskas, A.J.: Integration of renewable deep eutectic solvents with engineered biomass to achieve a closed-loop biorefinery. Proc. Natl. Acad. Sci. U.S.A. **116**(28), 13816–13824 (2019). https://doi.org/10.1073/pnas.1901596116
44. Wang, Y., Fan, C., Hu, H., Li, Y., Sun, D., Wang, Y., Peng, L.: Genetic modification of plant cell walls to enhance biomass yield and biofuel production in bioenergy crops. Biotechnol. Adv. **34**(8), 997–1017 (2016). https://doi.org/10.1016/j.biotechadv.2016.06.001
45. Yoo, C.G., Pu, Y., Ragauskas, A.J.: Ionic liquids: promising green solvents for lignocellulosic biomass utilization. Curr. Opin. Green Sustain. Chem. **5**, 5–11 (2017). https://doi.org/10.1016/j.cogsc.2017.03.002
46. Agrawal, R., Verma, A., Satlewal, A.: Application of nanoparticle-immobilized thermostable β-glucosidase for improving the sugarcane juice properties. Innov. Food Sci. Emerg. Technol. **33**, 472–482 (2016). https://doi.org/10.1016/j.ifset.2015.10.020
47. Gunny, A.A.N., Arbain, D.: Ionic liquids: green solvent for pretreatment of lingnocellulosic biomass. Adv. Mater. Res. **701**, 399–402 (2013). https://doi.org/10.4028/www.scientific.net/amr.701.399
48. Lim, W., Gunny, A.A.N., Kasim, F.H., AlNashef, I.M., Arbain, D.: Alkaline deep eutectic solvent: a novel green solvent for lignocellulose pulping. Cellulose **26**(6), 4085–4098 (2019). https://doi.org/10.1007/s10570-019-02346-8.
49. Li, P., Yang, C., Jiang, Z., Jin, Y., Wu, W.: Lignocellulose pretreatment by deep eutectic solvents and related technologies: a review. J. Bioresour. Bioprod. **8**(1), 33–44 (2023). https://doi.org/10.1016/j.jobab.2022.11.004
50. Sumiati, T., Suryadi, H., Harmita, Sutriyo.: Comparison of the deep euteutic solvent (DES) solvent for extracting lignin from the lignocellulosic material of pineapple leaves. Pharmacogn. J. **13**(6), 1702–1709 (2021). https://doi.org/10.5530/pj.2021.13.219
51. Satlewal, A., Agrawal, R., Bhagia, S., Das, P., Ragauskas, A.J.: Rice straw as a feedstock for biofuels: availability, recalcitrance, and chemical properties. Biofuels Bioprod. Biorefin. **12**(1), 83–107 (2018). https://doi.org/10.1002/bbb.1831
52. Liu, X., Wei, W., Wu, S.: Synergism of organic acid and deep eutectic solvents pretreatment for the co-production of oligosaccharides and enhancing enzymatic saccharification. Biores. Technol. **290**, 121775 (2019). https://doi.org/10.1016/j.biortech.2019.12177

53. Tian, D., Guo, Y., Hu, J., Yang, G., Zhang, J., Luo, L., Xiao, Y., Deng, S., Deng, O., Zhou, W., Shen, F.: Acidic deep eutectic solvents pretreatment for selective lignocellulosic biomass fractionation with enhanced cellulose reactivity. Int. J. Biol. Macromol. **142**, 288–297 (2020). https://doi.org/10.1016/j.ijbiomac.2019.09.100
54. Ma, H., Fu, P., Zhao, J., Lin, X., Wu, W., Yu, Z., Xia, C., Wang, Q., Gao, M., Zhou, J.: Pretreatment of wheat straw lignocelluloses by deep eutectic solvent for lignin extraction. Molecules **27**(22), 7955 (2022). https://doi.org/10.3390/molecules27227955.PMID:36432056;PMCID: PMC9697946
55. Ahorsu, R., Constanti, M., De María, P.D., Medina, F.: Synergy of ball milling, microwave irradiation, and deep eutectic solvents for a rapid and selective delignification: walnut shells as model for lignin-enriched recalcitrant biomass. Biomass Convers. Biorefinery **14**(11), 11849–11861 (2022). https://doi.org/10.1007/s13399-022-03566-y
56. Jamaldheen, S.B., Kurade, M.B., Basak, B., Yoo, C.G., Oh, K.K., Jeon, B., Kim, T.H.: A review on physico-chemical delignification as a pretreatment of lignocellulosic biomass for enhanced bioconversion. Biores. Technol. **346**, 126591 (2022). https://doi.org/10.1016/j.biortech.2021.126591
57. Gunny, A.A.N., Arbain, D., Nashef, E.M., Jamal, P.: Applicability evaluation of deep eutectic solvents–cellulase system for lignocellulose hydrolysis. Bioresour. Technol. **181**, 297–302 (2015). https://doi.org/10.1016/j.biortech.2015.01.057.
58. Liu, Y., Zheng, X., Tao, S., Hu, L., Zhang, X., Lin, X.: Process optimization for deep eutectic solvent pretreatment and enzymatic hydrolysis of sugar cane bagasse for cellulosic ethanol fermentation. Renew. Energy **177**, 259–267 (2021). https://doi.org/10.1016/j.renene.2021.05.131
59. Kim, K.H., Yoo, C.G.: Challenges and perspective of recent biomass pretreatment solvents. Front. Chem. Eng. **3** (2021). https://doi.org/10.3389/fceng.2021.785709
60. Wang, W., Lee, D.: Lignocellulosic biomass pretreatment by deep eutectic solvents on lignin extraction and saccharification enhancement: a review. Biores. Technol. **339**, 125587 (2021). https://doi.org/10.1016/j.biortech.2021.125587
61. Liu, Y., Chen, W., Xia, Q., Guo, B., Wang, Q., Liu, S., Liu, Y., Li, J., Yu, H.: Efficient cleavage of lignin-carbohydrate complexes and ultrafast extraction of lignin oligomers from wood biomass by microwave-assisted treatment with deep eutectic solvent. Chemsuschem **10**(8), 1692–1700 (2017). https://doi.org/10.1002/cssc.201601795
62. Scelsi, E., Angelini, A., Pastore, C.: Deep eutectic solvents for the valorisation of lignocellulosic biomasses towards fine chemicals. Biomass **1**(1), 29–59 (2021). https://doi.org/10.3390/biomass1010003
63. Huang, Z., Feng, G., Lin, K., Pu, F., Tan, Y., Tu, W., Han, Y., Hou, X., Zhang, H., Zhang, Y.: Significant boost in xylose yield and enhanced economic value with one-pot process using deep eutectic solvent for the pretreatment and saccharification of rice straw. Ind. Crops Prod. **152**, 112515 (2020). https://doi.org/10.1016/j.indcrop.2020.112515

Deep Eutectic Solvent (DES) and Ionic Liquid (IL) in Biodiesel Production

Sarina Sulaiman, Dzun Noraini Jimat, Siti Hajar Yusoff, and Harumi Veny

Abstract This study reviews the advancement of deep eutectic solvent (DES) in biodiesel production. Biodiesel or better-known as FAME is fatty acid alkyl ester derived from renewable resources such as plant waste, animal fats, vegetable oils and lipid of microorganisms. Furthermore, glycerol removal in biodiesel production using ionic liquid i-s becoming popular and need to be explored. Ionic liquid compared to many solvents and catalysts offers many advantages such as low vapor pressure, highly soluble in organic and inorganic solvent, high catalytic activity, excellent thermal stabilities, and possibility on desired design. The study will also focus on the deep eutectic solvent (DES) and the most used solvents in biodiesel processes were DESs based on choline chloride and ethylene glycol or glycerol as hydrogen bond donors. Finally, the possibility of DES recycling and reuse is also emphasized, showing bright prospects for the application of DESs in biodiesel production. DES shows a significant contribution towards the improvement of the biodiesel production.

1 Introduction

The demand for biodiesel has constantly increased throughout last decade due to rising oil prices and environmental concerns. World population growth is increasing and the use of petroleum derived energy resulting in high emissions of pollutants into the atmosphere. According to Merchant Research and United Nations Conference on Trade and Development (2022), the global biodiesel production is expected to

S. Sulaiman (✉) · D. N. Jimat
Department of Chemical Engineering and Sustainability, Faculty of Engineering, International Islamic University Malaysia (IIUM), Kuala Lumpur, Malaysia
e-mail: sarina@iium.edu.my

S. H. Yusoff
Department of Electrical and Computer Engineering, Faculty of Engineering, International Islamic University Malaysia, Kuala Lumpur, Malaysia

H. Veny
Faculty of Chemical Engineering, Universiti Teknologi MARA (UiTM), Shah Alam, Selangor, Malaysia

D. N. Jimat (ed.), *Deep Eutectic Solvents*, Engineering Materials,
https://doi.org/10.1007/978-981-95-1912-5_8

be 30% higher than 2019 level in 2023–2025 where Europe is the main biodiesel producing and consuming region with 40% followed by Asia Pacific and America as shown in Fig. 1.

Biodiesel or better-known FAME is fatty acid alkyl ester derived from renewable resources such as plant waste, animal fats, vegetable oils and lipid of microorganisms. Since feedstock for biodiesel comes from renewable sources, the availability of it is very much dependent on geographical factor where the biodiesel is manufactured. Countries with tropical climate such Malaysia and Indonesia use plants that easily grow in this climate such as palm oil. But mass consumption of palm oil for energy development has raised an issue whereby this sector must compete with food production as both use the same feedstock. Therefore, research have shifted to finding other alternatives to substitute palm oil as feedstock.

Studies have brought findings of utilization of waste cooking oil as feedstock. Not only it helps to reduce land and water pollution, but utilization of waste cooking oil also cuts much cost in production due to availability of the waste. Study done by group of University of Science Malaysia's (USM) students showed that the majority of social community failed to manage their waste cooking oil, rather the waste are discarded into inappropriate channels such as drain, soil and dustbin. Therefore, using waste cooking oil as feedstock in biodiesel production reduces problems related to waste management.

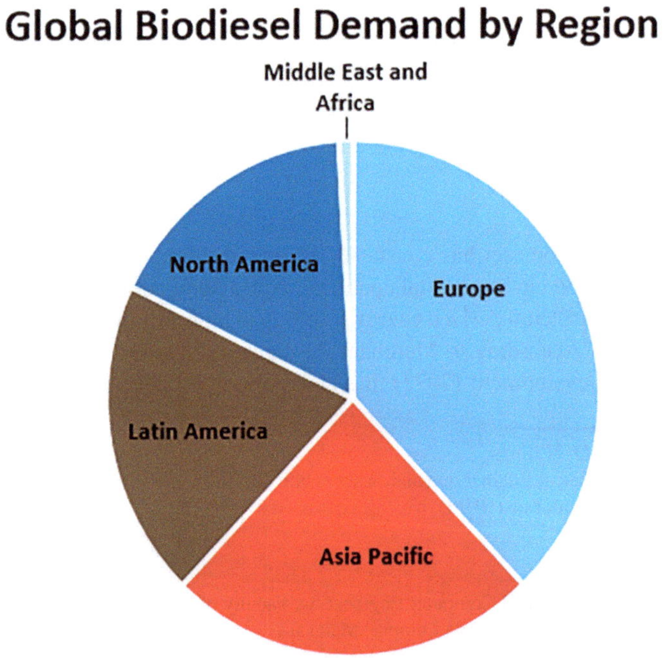

Fig. 1 Global biodiesel demand by region (Merchant research, 2024)

Jatropha *curcas* is becoming more well-known among biodiesel producers since the plant can be an ideal substitute to palm oil as it is not used in food production. Physically, Jatropha possesses the similar fat content as those in palm oil, in which it can be extracted for biodiesel production. In other countries such as the US, corn is the main feedstock while India and Brazil use soybeans as raw material for synthesis of biodiesel [1].

2 Process Technology

Of all processes technology for biodiesel production, two steps catalysis by far is the most popular technique used in the conversion due to its simplicity and easy recovery of product. Using this method, oil is esterified in the presence of catalyst to reduce the moisture content to a level that inhibits saponification [2]. In the second step oil is subject to transesterification process to form biodiesel [3]. The majority of researchers agree that in the transesterification reaction, alkaline catalyst is more favorable as compared to acid catalyst due to high catalytic activity [4].

Canaki and Van Gerpen in their study showed that oil, especially low cost oil such as waste cooking oil, usually is converted into biodiesel in two steps reaction to reduce high fatty acid content to reusable ester [5] Pretreatment of low-cost oil in biodiesel production is important to avoid soap formation or otherwise cause deactivation of catalyst [6]. On the other hand, researchers employ supercritical method, microwave, and ultrasonication method in biodiesel production in the absence of catalyst [7]

Lin and collogues demonstrated that employing microwave in biodiesel production, complete conversion in ionic liquid took only 6 min [8] compared to those using conventional method which take at least 1 h of reaction time. As revealed by Thanh et al. in their review, exposing mixture for biodiesel production to ultrasonic irradiation can speed up the reaction time up to 95% of those required in normal process [9]. These two methods undeniably improve reaction timing exponentially but in large scale of biodiesel production, they are less likely to be employed due to economical reason.

Based on huge number of studies done, alcohol is identified as the most frequently used organic solvent in the reaction; particularly methanol due to low cost and availability of the liquid [10]. Nevertheless, other alcohol such as ethanol, butanol, and propanol are likely to substitute methanol in reactions that favor alcohol with long carbon chain. Besides, the tendency of methanol to deactivate enzyme activity initiates researchers to change to other solvent for biodiesel production [11].

3 Catalyst Selection

Similar to process technology, there are a number of choices from which one can select type of catalyst to be used in biodiesel production. Depending on the chemical, physical and biological content of the feedstock, the requirement for preliminary treatment, catalyst selection and mode of processes also vary. Famous selection of catalyst constitutes acid catalyst, basic, homogeneous, heterogeneous, and biocatalyst [12]. These groups of catalysts can be distinctively disguised in terms of their performance with various feedstock and reaction solvent.

Acid catalyst favors conversion of microalgae and oils with high free fatty acid (FFA) content into biodiesel [13] while basic catalyst such as sodium and potassium hydroxide are preferable due to their higher catalytic activity it possesses in biodiesel synthesis. Regardless of various basic and acidic catalysts been used in the process, there are still some limitations caused by catalytic structure and properties that inhibit reaction.

Drawbacks induced by basic and acid catalysts lead to utilization of biocatalyst to manufacture biodiesel. Biocatalyst or better known as enzyme, compared to others is highly reusable and can be recycled for a few times without losing its efficiency. Besides easy product recovery, reaction using enzyme catalyst gives highest yield in standard condition. Figure 2 illustrates the overall process flow of DES and Ionic liquid that works as solvent or catalyst in biodiesel production.

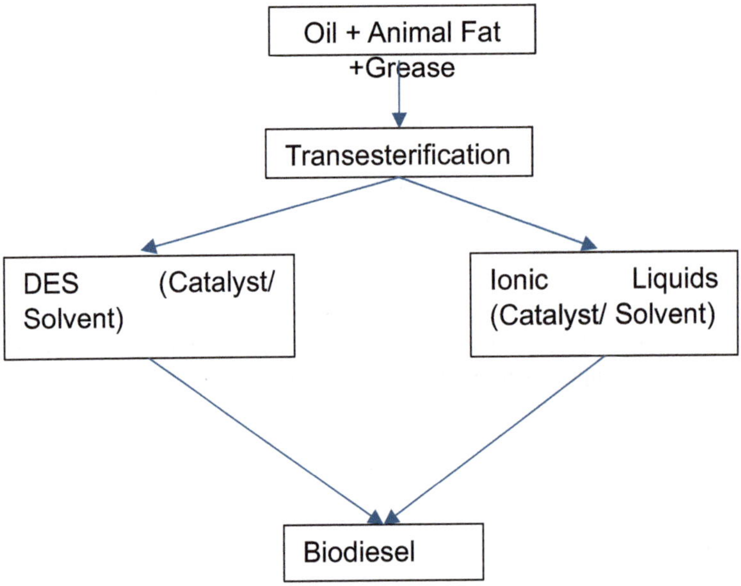

Fig. 2 Overall process to produce biodiesel using ionic liquid and DES

3.1 Ionic Liquid as Catalyst, Solvent, and Co-Solvent

Regardless of some advantages biocatalyst offers, it also exhibits several drawbacks in transesterification application. Major weaknesses is that enzyme is quite costly relative to other catalysts [14]. Besides, enzyme can be easily deactivated due to blockage in the active sites and requirement for complicated separation process as well as the potential to increase environmental concern due to its toxicity [15].

Due to that reason, recent years have witnessed that studies on the same theme has shifted to finding more feasible replacement of acid catalyst, basic and biocatalyst. Research has brought to using ionic liquid as the potential substitute of all type of catalyst. Function of ionic liquid is not limited to as catalyst; it serves as to act as solvent and co-solvent in biodiesel production to replace the most frequently used alcohol such as methanol and ethanol.

Ionic liquid is generally described as an organic salt that constitutes cations and anions that exist as liquid at room temperature. Modern technology has enabled more and more groups of ionic liquid to be developed by customized synthesis, all of which are useful in chemical synthesis in broad applications. In biofuel production, there are a number of remarkable findings related to the association of ionic liquid and biodiesel, but there is still a wide gap in which more studies and research are required to fill it in.

As mentioned by Han et al. [16], ionic liquid has low flammability and low vapor pressure, high ionic and thermal conductivity, high dissolution capability towards many substrates, non-corrosive, non-toxic, non-volatile and is a stable at wide range of temperature. Chemically, ionic liquids is more favorable as compared to other substances because its properties such as polarity and hydrophobicity can be finely tuned by customized selection of cations and anions [15].

Earle et al. in their study divided ionic liquid into three major categories; Brønsted acidic ionic liquid, Brønsted base, and Lewis acid ionic liquid [17]. In contrast, Ratti categorized ionic liquid into two major groups, simple salts and binary ionic liquid [18]. Being differ to these groups, Andreani and Rocha reviewed that ionic liquid can be clustered into two major groups, one is based on chlorometallate anions and another that does not contain metals [10]. Either way of grouping, as the name implies, these groups of ionic liquids are differ in terms of content in the liquid.

As per current trend in biodiesel production, focus is on finding ionic liquid that can serve as catalyst and the right proportion of it as well as solvent with highest reproducibility. Except for a few, attempts to test catalytic activity of ionic liquid and performance as a solvent are done separately. It means that there is a very minimum discovery on how ionic liquid can enhance biodiesel production by performing the dual functions.

The following reviews several studies involving some ionic liquid in catalysis and solvent in biodiesel production. Comparative study of bio-based ionic liquid in methanol by Reddy et al. has brought to a finding that choline hydroxide is more efficient as catalyst than that choline imidazole with 95 ± 1% yield of biodiesel from *Jatropha curcas* oil at 60°C in 4 h reaction time [18]. On top of high catalytic performance, they proved that catalyst is recyclable for at least four times without losing its catalytic activity.

In different study of Brønsted acidic ionic liquid, 1-(3-sulfonic acid)propyl-3–3-methylimidazole hydrosulfate, $[HO_3S\text{-pmim}]H_2SO_4$ exhibits the highest catalytic activity with 93% biodiesel yield in 8 h at 120°C and waste oil and methanol ratio of 1:12. The comparison made is between several other imidazolium ionic liquid such as $[Hmim]HSO_4$, $[Hmim]H_2PO_4$, $[Hmim]BF_4$, $[C4mim]HSO_4$, $[C4mim]H_2PO_4$, $[C4mim]BF_4$, $[HO_3S\text{-pmim}]HSO_4$, $[HO_3S\text{-pmim}]BF_4$, and $[HO_3S\text{-pmim}]H_2PO_4$. Similarly, the catalyst is reusable for more than four times without losing much of its efficiency [19].

Another group of ionic liquid that has broad application in many chemical reactions is cholinium-based ionic liquid. In biodiesel production, Isahak and colleagues worked on a number of choline salts using different group of metal namely choline chloride.$ZnCl_5^-$, and choline chloride. $FeCl_7^-$ to determine their catalytic activities. At the end of the study they concluded that after 4 h of reaction time at 65 °C choline chloride. $ZnCl_5^-$ showed a better performance of catalysis [20]. Using this ionic liquid biodiesel yield is 92% compared 89.5% when using choline chloride $FeCl_7^-$.

Imidazolium-based material is a group of ionic liquid that has huge number application in industry with no exception in biodiesel production. In conversion of soybeans to biodiesel using alcoholysis method, 1-butyl-3-methyl morphine hydroxide ([Hnmm]OH) demonstrated similar to cholinium-based ionic liquid, that is high reusability. The liquid is recyclable with biodiesel yield more than 90% even after five times of reaction for 1.5 h at 70°C [21]

3.2 Deep Eutectic Solvent Des

DES or deep eutectic solvent is getting noticed due to the problems associated with the industrial purification method for biodiesel where wet washing of biodiesel consumes a large amount of fresh water. The wastewater after biodiesel purification must be cleaned before it is discharged into the environment, which requires a large amount of energy. Therefore, more environmentally friendly methods for the purification of biodiesel are necessary for greener and environmentally friendly processing. Liquid–liquid extraction with deep eutectic solvents (DESs) is one of the alternative methods to remove glycerol from biodiesel after the transesterification. Furthermore green solvents such as DES have high thermal stability and low volatility.

DES is done by mixing the hydrogen bond donor (HBD) and hydrogen bond acceptor (HBA). Hydrogen bond donor are small, neutral, covalently bonded molecules which are capable of hydrogen bonding, such as carboxylic acids, polyols, and sugars. The hydrogen bond acceptor are normally ionic compounds, generally quaternary ammonium salts as shown in Table 1.

This table shows the basic components used in the formation of DES, with HBD molecules donating hydrogen bonds and HBA molecules accepting them to form a eutectic mixture. DESs are highly influential in biodiesel production and are regarded

Table 1 Components used in the formation of DES

Component	Type	Examples
Hydrogen bond donor (HBD)	Small, neutral, covalently bonded molecules capable of hydrogen bonding	Carboxylic acids (e.g., acetic acid), polyols (e.g., glycerol), sugars (e.g., glucose)
Hydrogen bond acceptor (HBA)	Ionic compounds, typically quaternary ammonium salts	Quaternary ammonium salts (e.g., choline chloride), metal halides (e.g., zinc chloride)

as greener counterparts to Ionic Liquids (ILs). Their desirable properties, such as excellent thermal and chemical stability, non-corrosive nature, and minimal vapor pressure, make them a sustainable choice.

4 DES in Biodiesel

DES or deep eutectic solvent is getting noticed due to the problems associated with the industrial purification method for biodiesel where wet washing of biodiesel consumes a large amount of fresh water. The wastewater after biodiesel purification must be cleaned before it is discharged into the environment, which requires a large amount of energy. Therefore, more environmentally friendly methods for the purification of biodiesel are necessary for greener and environmentally friendly processing. Liquid–liquid extraction with deep eutectic solvents (DESs) is one of the alternative methods to remove glycerol from biodiesel after the transesterification. Furthermore, green solvents such as DES have high thermal stability and low volatility.

Potchamyou Ngatcha et al. [22] reported that the catalytic activity of DESs in the transesterification of microalgae lipids into biodiesel, offers a safer alternative to the toxic and corrosive catalyst H_2SO_4 [22]. DES ChCl–$SnCl_2$ (choline chloride–tin II chloride) showed higher catalytic activity in transesterification due to its higher acidity compared to DES ChCl–$ZnCl_2$ (choline chloridezinc chloride). The fatty acid methyl ester (FAME) conversion from microalgae lipid was 36.75 mg g^{-1} under ideal conditions (6 molar ratio methanol–lipid with 8 vol% DES dosage in methanol at 140 °C for 420 min). The DES catalyst (ChCl–$SnCl_2$) drove the reaction chemically and did not show mass transfer limitation.

Khan et al. [23] produced biodiesel by transesterification with a deep eutectic solvent (DES), without requiring extra washing, using waste cooking oil and grease from an on-campus dining facility [23]. Biodiesel synthesized using DES was reported to be the blend of saturated and unsaturated FAMEs and meet viscosity standards under normal climate temperatures (Table 2).

Manurung et al. [24] studied the effectiveness of Deep Eutectic Solvent (DES) based choline chloride: 1,2 propanediol as co-solvent in biodiesel purification. DES used on the basis of Choline Chloride and 1,2-Propanediol with a 1: 2 and 1: 3 molar ratio were applied to purification of biodiesel at concentration of 1,5; 10; and 15% (w/

Table 2 Summary of the parameters using DES in biodiesel production

Raw material	DES	Conversion	Parameter	References
Microalgae lipids	ChCl–$ZnCl_2$	36.75 mg g^{-1}	– 6 molar ratio methanol–lipid – 8 vol% DES dosage in methanol – 140 °C for 420 min	Potchamyou Ngatcha et al. [22]
Waste cooking oil and grease	Choline chloride/ glycerol	B30	– 600 g of waste cooking oil and grease was mixed with 120 g of methanol and heated to 65 °C	Khan et al. [23]
Degummed palm oil (DPO)	Choline Chloride (salt) and 1,2-Propanediol	Purity of biodiesel was 98,88%	– Stirring: 300 rpm, Temperature: 65 C for 1 h – molar ratio of methanol: DPO of 6: 1, ethanol: DPO of 9:1 – catalyst concentration: 1% for methanol and 1.2% for ethanol	Manurung et al. [24]
Jatropha curcas oil	K_2CO_3/glycerol	98.88%	– 1:32.58 mol ratio of K_2CO_3/ glycerol, 8.96% w/w concentration of DES, and 69.58 min	Abdurrahman et al. [25]

w). It can be concluded that DES based ChCl / 1,2 Propanediol has good properties, and the yield of biodiesel was 98.88% at ethanol to oil molar ratio of 9: 1 and 5% DES in ethanolysis process [24].

Abdurrahman et al. [25] synthesized deep eutectic solvent as catalyst from different mole ratios of K_2CO_3/glycerol to produce biodiesel from Jatropha curcas oil. Optimum fatty acid methyl ester (FAME) yield of 98.28% was achieved at optimum conditions of 1:32.58 mol ratio of K_2CO_3/glycerol, 8.96% w/w concentration of DES, and 69.58 min. DES-based solvent demonstrated favourable properties with ASTM D6751-12 standard.

The studies highlight the potential of DES in biodiesel production but the focus on differs on catalysis. While both approaches achieved high biodiesel yields, their scalability and economic feasibility remain critical challenges. The use of DES, in both cases, aligns with principles of green chemistry, but more emphasis is needed on recyclability, cost analysis, and long-term performance.

4.1 DES Recycling and Reuse

Recyclability of deep eutectic solvents (DESs) involves their separation, purification, and reuse through sequential steps [26]. Due to their higher cost compared to organic solvents, recovering and recycling DESs enhance economically, where hydrophobic DESs being particularly advantageous for chemical reactions. Both hydrogen bond donor (HBD) and hydrogen bond acceptor (HBA) combination of DESs strongly affects recycling. Biphasic systems formed by the addition of hydrophilic DESs with potassium salt or sucrose facilitate phase separation and enables DESs to be recycled and recovered, and membrane separation enhances the separation of the hydrophilic DES. There is no change in the catalyst activity even after being used several times.

5 Conclusion

In conclusion, this study highlights the advancements in utilizing deep eutectic solvents (DESs) for biodiesel production, underscoring their significant contribution to improving the process. Biodiesel, derived from renewable resources such as plant waste, animal fats, vegetable oils, and microbial lipids, benefits from the enhanced glycerol removal capabilities offered by DESs and ionic liquids. While ionic liquids provide advantages like low vapor pressure, high solubility, excellent thermal stability, and customizable properties, DESs emerge as a greener alternative, particularly those based on choline chloride and hydrogen bond donors such as ethylene glycol or glycerol. Furthermore, the potential for recycling and reusing DESs demonstrates promising prospects for sustainable and cost-effective biodiesel production. This study reaffirms the critical role of DESs in advancing biodiesel technology and paving the way for more efficient and environmentally friendly energy solutions.

References

1. Biofuels Engineering Process Technology.: Focus on catalysts **2008**(12) (2008). https://doi.org/10.1016/s1351-4180(08)70584-1
2. Vicente, G., Martínez, M., Aracil, J.: Integrated biodiesel production: a comparison of different homogeneous catalysts systems. Bioresour. Technol. **92**(3) (2004). https://doi.org/10.1016/j.biortech.2003.08.014
3. Ullah, Z., Bustam, M.A., Man, Z.: Biodiesel production from waste cooking oil by acidic ionic liquid as a catalyst. Renew. Energy **77** (2015). https://doi.org/10.1016/j.renene.2014.12.040

4. Gnanaprakasam, A., Sivakumar, V.M., Surendhar, A., Thirumarimurugan, M., Kannadasan, T.: Recent strategy of biodiesel production from waste cooking oil and process influencing parameters: a review. J. Energy **2013** (2013). https://doi.org/10.1155/2013/926392
5. Canakci, M., Van Gerpen, J.: Biodiesel production from oils and fats with high free fatty acids. Trans. Am. Soc. Agric. Eng. **44**(6) (2001)
6. Dholakiya, B.Z.: Super phosphoric acid catalyzed biodiesel production from low cost feed stock. Arch. Appl. Sci. Res. Appl. Sci. Res. **4**(1) (2012)
7. Helwani, Z., Othman, M.R., Aziz, N., Kim, J., Fernando, W.J.N.: Solid heterogeneous catalysts for transesterification of triglycerides with methanol: a review. Appl. Catal. A: Gen. **363**(1–2) (2009). https://doi.org/10.1016/j.apcata.2009.05.021
8. Lin, Y.C., Yang, P.M., Chen, S.C., Lin, J.F.: Improving biodiesel yields from waste cooking oil using ionic liquids as catalysts with a microwave heating system. Fuel Process. Technol. **115** (2013). https://doi.org/10.1016/j.fuproc.2013.04.004
9. Georgogianni, K.G., Kontominas, M.G., Avlonitis, D., Gergis, V.: Transesterification of sunflower seed oil for the production of biodiesel: effect of catalyst concentration and ultrasonication. WSEAS Trans. Environ. Dev. **2**(2) (2006)
10. Andreani, L., Rocha, J.D.: Use of ionic liquids in biodiesel production: a review. Braz. J. Chem. Eng. **29**(1) (2012). https://doi.org/10.1590/S0104-66322012000100001
11. Shimada, Y., Watanabe, Y., Sugihara, A., Tominaga, Y.: Enzymatic alcoholysis for biodiesel fuel production and application of the reaction to oil processing. J. Mol. Catal.-B Enzym. **17**(3–5) (2002). https://doi.org/10.1016/S1381-1177(02)00020-6
12. Lam, M.K., Lee, K.T., Mohamed, A.R.: Homogeneous, heterogeneous and enzymatic catalysis for transesterification of high free fatty acid oil (waste cooking oil) to biodiesel: a review. Biotechnol. Adv. **28**(4), 500–518 (2010). https://doi.org/10.1016/j.biotechadv.2010.03.002
13. Sani, Y.M., Daud, W.M.A.W., Abdul Aziz, A.R.: Activity of solid acid catalysts for biodiesel production: a critical review. Appl. Catal. A: Gen. **470** (2014). https://doi.org/10.1016/j.apcata.2013.10.052
14. Fjerbaek, L., Christensen, K.V., Norddahl, B.: A review of the current state of biodiesel production using enzymatic transesterification. Biotechnol. Bioeng. **102**(5) (2009). https://doi.org/10.1002/bit.22256
15. Zhao, H., Baker, G.A.: Ionic liquids and deep eutectic solvents for biodiesel synthesis: a review. J. Chem. Technol. Biotechnol. **88**(1) (2013). https://doi.org/10.1002/jctb.3935
16. Han, D., Tang, B., Ri Lee, Y., Ho Row, K.: Application of ionic liquid in liquid phase microextraction technology. J. Sep. Sci. **35**(21) (2012). https://doi.org/10.1002/jssc.201200486
17. Earle, M.J., Plechkova, N.V., Seddon, K.R.: Green synthesis of biodiesel using ionic liquids. Pure Appl. Chem. **81**(11) (2009). https://doi.org/10.1351/PAC-CON-08-11-07
18. Ratti, R.: Ionic liquids: synthesis and applications in catalysis. Adv. Chem. **2014** (2014). https://doi.org/10.1155/2014/729842
19. Liu, S., Wang, Z., Yu, S., Xie, C.: Transesterification of waste oil to biodiesel using brønsted acid ionic liquid as catalyst. Bull. Chem. Soc. Ethiop. **27**(2) (2013). https://doi.org/10.4314/bcse.v27i2.14
20. Isahak, W.N.R.W., Ismail, M., Jahim, J.M., Salimon, J., & Yarmo, M. A. (2011). Transesterification of palm oil by using ionic liquids as a new potential catalyst. Trends Appl. Sci. Res. **6**(9). https://doi.org/10.3923/tasr.2011.1055.1062
21. Ren, Q., Zuo, T., Pan, J., Chen, C., Li, W.: Preparation of biodiesel from soybean catalyzed by basic ionic liquids [Hnmm]OH. Materials **7**(12) (2014). https://doi.org/10.3390/ma7128012
22. Potchamyou Ngatcha, A.D., Zhao, A., Zhang, S., Xiong, W., Sarker, M., Xu, J., Alam, M.A.: Determination of active sites on the synthesis of novel Lewis acidic deep eutectic solvent catalysts and kinetic studies in microalgal biodiesel production. RSC Adv. **13**(15) (2023). https://doi.org/10.1039/d3ra00543g
23. Khan, N., Park, S.H., Kadima, L., Bourdeau, C., Calina, E., Edmunds, C.W., Pursell, D.P.: Locally sustainable biodiesel production from waste cooking oil and grease using a deep eutectic solvent: characterization, thermal properties, and blend performance. ACS Omega **6**(13) (2021). https://doi.org/10.1021/acsomega.1c00556

24. Manurung, R., Syahputra, A., Alhamdi, M.A.: Purification of palm biodiesel using deep eutectic solvent (DES) based choline chloride (ChCl) and 1,2-Propanediol ($C_3H_8O_2$). J. Phys.: Conf. Ser. **1028**(1) (2018). https://doi.org/10.1088/1742-6596/1028/1/012032
25. Abdurrahman, A., Waziri, S.M., Ajayi, O.A., Dabai, F.N.: Optimization of potassium carbonate-based DES as catalyst in the production of biodiesel via transesterification. J. Niger. Soc. Phys. Sci. **5**(1) (2023). https://doi.org/10.46481/jnsps.2023.1048
26. Maheshwari, P., Haider, M.B., Yusuf, M., Klemeš, J.J., Bokhari, A., Beg, M., Al-Othman, A., Kumar, R., Jaiswal, A.K.: A review on latest trends in cleaner biodiesel production: role of feedstock, production methods, and catalysts. J. Clean. Prod. **355** (2022). https://doi.org/10.1016/j.jclepro.2022.131588

Challenges, Toxicity and Socioeconomic and Environmental Impact Using Des

Dzun Noraini Jimat and Muhammad Hasya Hasnuddin

Abstract This chapter covers the challenges of usages of deep eutectic solvents, toxicity and socioeconomic and environmental impacts. Deep eutectic solvents (DESs) are emerging green solvents that are widely used in various applications, such as organic synthesis, electrochemistry, extraction of active compounds, synthesis and separation processes, solubility enhancement of hydrophobic drugs, and drug delivery formulations. Thus, the used DES must be well characterized not only its physicochemical properties but also cytotoxicity to ensure its implementation in real industry processes competitive, safer, and eco-friendly solvents compared to conventional solvents. All users either in research and developments of industry must have regular safety assessment of handling practices and disposal methods of DES aligned with sustainable development gold and circular economy. Furthermore, the long-term environmental impact needs to be studied when DES becomes widely used, as there is a possibility of bioaccumulation and the impact of breakdown products of DES in our environment. Regulations and guidelines for DES in all applications should be established to ensure their safe use and sustainability, especially in large-scale applications.

1 Challenges of Usages of Deep Eutectic Solvents (DESs) in Industry

Deep eutectic solvents (DESs) are emerging green solvents that are widely used in various applications, such as organic synthesis, electrochemistry, extraction of active compounds, synthesis and separation processes, solubility enhancement of hydrophobic drugs, and drug delivery formulations [1]. Figure 1 showed the primary features of solvents that preferable used in real industry processes. Thus, the used DES must be well characterized not only its physicochemical properties but also cytotoxicity to ensure its implementation in real industry processes competitive, safer, and

D. N. Jimat (✉) · M. H. Hasnuddin
Department of Chemical Engineering and Sustainability, Kulliyyah of Engineering, International Islamic University Malaysia (IIUM), Gombak, Kuala Lumpur, Malaysia
e-mail: jnoraini@iium.edu.my

D. N. Jimat (ed.), *Deep Eutectic Solvents*, Engineering Materials,
https://doi.org/10.1007/978-981-95-1912-5_9

eco-friendly solvents compared to conventional solvents. All users either in research and developments of industry must have regular safety assessment of handling practices and disposal methods of DES aligned with sustainable development gold and circular economy.

However, further study is necessary to understand their adaptability and selectivity for diverse applications. Understanding their molecular interactions and mechanisms is essential to ensure their effectiveness and performance compared to conventional solvents. Setting optimal operating conditions, such as temperature and DES molar ratio, with adequate agitation is of utmost importance in processes such as the extraction of active compounds and chemical reactions. The occurrence of mass transfer is influenced by the physicochemical properties of the chosen DES, such as its density, polarity, and volatility.

It is also important to set up a pilot-scale process to mimic industrial-scale operations, as a chemical reactor constructed at the lab scale is typically made of glass, which has different heating rates and heat transfer properties compared to a large-scale chemical reactor made from stainless steel. Additionally, controlling the optimal

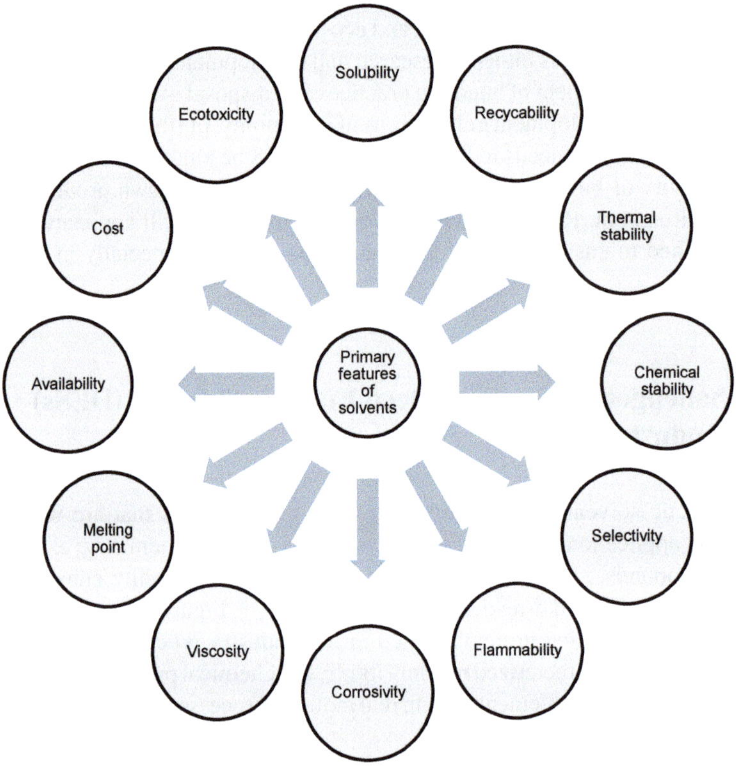

Fig. 1 Primary features of solvents used in any applications

conditions in a large reactor presents different challenges. Thus, a pilot-scale chemical reactor can help connection the gap between lab-scale experiments and full-scale industrial applications by allowing researchers to optimize process parameters, evaluate scalability, and identify potential operational challenges before large-scale execution.

Furthermore, its scalability and cost-effectiveness also need to be considered before it can be implemented in the industry. Several factors must be evaluated, such as the variability of DES physicochemical properties, as there may be restrictions on its large-scale consumption. Additionally, the cost of raw materials, installation of new equipment, and energy utilization (including pumping, heating, and stirring) must be taken into account, along with the management of side products and waste in both upstream and downstream operations [2]. The separation and purification of used DES present challenges, as it is crucial to select appropriate methods capable of removing most impurities while minimizing solvent degradation. Moreover, the recovery and reusability of DES are also challenging, as maintaining its performance over multiple usages cycles in any application or treatment process requires careful optimization. Figure 2 shows four main challenges of applying new solvents such as deep eutectic solvents as an alternative to conventional solvents in industry.

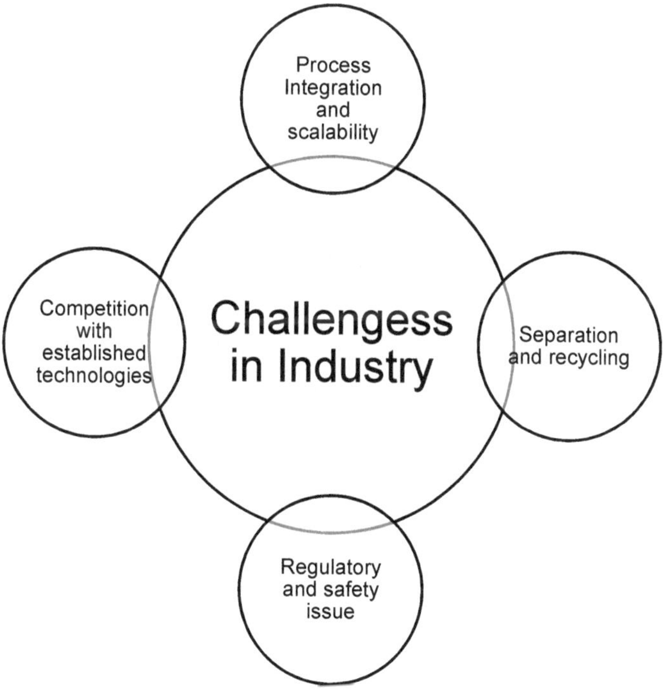

Fig. 2 Four main challenges of application of new solvents such as DESs in industry

2 Toxicity

One of the prime advantages of DESs is their toxicity itself; the majority of DESs are based on non-toxic, biodegradable materials such as choline chloride, urea, and glycerol, of natural origin, and safe for pharmaceutical, cosmetic, and even food applications [3, 4]. Since DES are derived either from natural or from biocompatible materials, they are assumed to be non-toxic or at least less hazardous than the conventional chemical solvents [5]. For example, choline chloride is a nutritional component and can be present in most living organisms, which further contributes to low toxicity of DES [6].

However, it was found that not every DES is completely harmless. Certain DES metal salts or even made of synthetic components may be toxic and therefore should be treated with care. For instance, such DES will have to be disposed of carefully to avoid soil or water contamination. The potential toxicity in some DES dictates that caution must be handled carefully when considering its safety, particularly when DES are handled in environments likely to expose them to either humans or animals. As noted by Maugeri and Leitner [7], despite these concerns, most DESs are safer and more sustainable than conventional solvents and hence herald a promising option toward greener chemistry [4, 8].

Lomba L. and co-authors reported that the cytotoxic effects of some DESs are influenced by different mixtures of HBAs or HBDs, the presence of organic acids as HBDs (which may alter the pH), the molar ratios of the components, the occurrence of synergistic toxicity, and the inclusion of components that change metabolic pathways or viscosity levels. Thus, their safety and environmental toxicity need to be thoroughly verified before utilization.

Choline chloride (ChCl) is the most common HBA which is a component of the vitamin B complex, so it intervenes in cellular metabolism [9]. Previous studies reported that calcium oxalate crystals were observed when the fish cell lines (CCO) were treated with ChCl/oxalate which can harm cells [10]. This result showed that this DES mixture possesses higher cytotoxicity towards these cell lines. The same authors reported that combination of CHCl with sugar (such as glucose, fructose, xylose) or organic acid (such as malic acid) exhibited low cytotoxicity [11].

Other studies prepared eight natural deep eutectic solvents (NADES) containing organic acid (malic acid or citric acid). They assessed the cytotoxicity of prepared NADES toward three human cell lines (HEK-293T, HeLa, and MCF-7 cells), and antioxidative activity was measured by the Oxygen Radical Absorbance Capacity (ORAC) method [12]. They found that all tested NADES containing carboxylic acid were almost safe and considered environmentally harmless.

The pH of DES mixtures rely on its HBD [1]. A highly acidity can cause denaturation of membrane protein and death of cells. A higher toxicity of the mixtures composed of organic acids may cause an alteration in the cellular proliferation and metabolic pathways [8]. The carbonated chain's length and the existence of certain functional groups determine pH variations (benzene appears to be less hazardous than carboxylic or alcohol groups) [13]. Molar ratio of DES components is another factors that influence its cytotoxicity [1]. A previous study reported that ChCl/urea at a 1:3 molar ratio and a 1:2 molar ratio exhibited acute toxicity (via in vitro MTT assay) and high toxicity (in vivo studies), respectively [14].

Zhao and co-researchers reported that the individual components of DESs were less toxic than their mixtures due to charge delocalization in hydrogen bonding [8]. This indicates a synergistic effect in certain DESs with HBDs, such as tartaric and citric acids, which have a more compact structure [8].

Previous studies observed that ChCl/tartaric acid (2:1), ChCl/citric acid (1:1), ChCl/malic acid (1:1), and ChCl/D-sorbitol (1:1) exhibited high viscosities with values > 10 Pa·s. The viscosities of these mixtures are higher due to the excess hydroxyl groups generates more hydrogen bonds and increases the attractive forces between molecules (H-bonding network, van der Waals forces, and electrostatic interplays) [8]. Dai and co-researchers demonstrated that the water content of DESs affects their viscosity and that the hydrogen bond structure of the DES components is disrupted if the water content exceeds 50% [1, 15]. A high viscosity of DES mixtures may lead to high toxicity. Studies have found that the inclusion of highly viscous substances (above 500 mPa·s) in a cell medium can cause significant changes in cytoplasmic viscosity (homeostatic between 1 and 400 mPa·s) [16], and could lead to cell death [17].

3 Socioeconomic and Environmental Impacts

Development and use of DES therefore have significant socio-economic benefits. Several industries are moving toward greener chemicals, and DES represents a cost-effective alternative to conventional solvents whose origin usually relies on resource-intensive, expensive, and non-renewable petroleum-based feedstocks [3, 4].

Because DES is biodegradable and non-toxic, it also lowers waste management costs and reduces the need for environmental cleanup. When companies use DES, compliance with environmental regulations is easier to achieve and thus cheaper. New avenues of innovation also come with DES. On the other hand, DES is especially promising in pharmaceutical applications, where it enables the safe formulation of drugs and bioactive compounds.

Switching to DES will also reduce dependence on petroleum, thereby cushioning industries from price fluctuations and supply disruptions. The shift toward sustainability by governments and businesses may also result in financial incentives and subsidies for the use of greener technologies such as DES, further encouraging their adoption.

DES can also be recovered and recycled thereby reducing waste and conserving resources. This aligns with the concept of the circular economy, where materials are kept in use for as long as possible to extract their maximum value while minimizing waste generation [18]. In addition to this, the long-term environmental impact needs to be studied when DES becomes widely used, as there is a possibility of bioaccumulation and the impact of breakdown products of DES in our environment. Regulations and guidelines for DES in all applications should be established to ensure the safe use and sustainability of DES, especially in large-scale applications.

References

1. Lomba, L., Ribate, M.P., Zaragoza, E., Concha, J., Garralaga, M.P., Errazquin, D., García, C.B., Giner, B.: Deep eutectic solvents: are they safe? Appl. Sci. (Switzerland) **11**(21) (2021). https://doi.org/10.3390/app112110061
2. Rente, D., Cvjetko Bubalo, M., Panić, M., Paiva, A., Caprin, B., Radojčić Redovniković, I., Duarte, A.R.C.: Review of deep eutectic systems from laboratory to industry, taking the application in the cosmetics industry as an example. J. Clean. Prod. **380**(February) (2022). https://doi.org/10.1016/j.jclepro.2022.135147
3. Abbott, A.P., Boothby, D., Capper, G., Davies, D.L., Rasheed, R.K.: Deep Eutectic solvents formed between choline chloride and carboxylic acids: versatile alternatives to ionic liquids. J. Am. Chem. Soc. **126**(29), 9142–9147 (2004). https://doi.org/10.1021/ja048266j
4. Smith, E.L., Abbott, A.P., Ryder, K.S.: 2014-SmithAbbottRyder-deep eutectic solvents (DESs) and their applications. Chem. Rev. **114**, 11060–11082 (2014)
5. Zhang, Q., De Oliveira Vigier, K., Royer, S., Jérôme, F.: Deep eutectic solvents: syntheses, properties and applications. Chem. Soc. Rev. **41**(21), 7108–7146 (2012). https://doi.org/10.1039/c2cs35178a
6. Francisco, M., Van Den Bruinhorst, A., Kroon, M.C.: New natural and renewable low transition temperature mixtures (LTTMs): screening as solvents for lignocellulosic biomass processing. Green Chem. **14**(8), 2153–2157 (2012). https://doi.org/10.1039/c2gc35660k
7. Maugeri, Z., Leitner, W., Domínguez De María, P.: Chymotrypsin-catalyzed peptide synthesis in deep eutectic solvents. Eur. J. Org. Chem. **20**, 4223–4228 (2013). https://doi.org/10.1002/ejoc.201300448
8. Zhao, B.Y., Xu, P., Yang, F.X., Wu, H., Zong, M.H., Lou, W.Y.: Biocompatible deep eutectic solvents based on choline chloride: characterization and application to the extraction of Rutin from Sophora Japonica. ACS Sustain. Chem. Eng. **3**(11), 2746–2755 (2015). https://doi.org/10.1021/acssuschemeng.5b00619
9. Florindo, C., Oliveira, F.S., Rebelo, L.P.N., Fernandes, A.M., Marrucho, I.M.: Insights into the synthesis and properties of deep eutectic solvents based on cholinium chloride and carboxylic acids. ACS Sustain. Chem. Eng. **2**(10), 2416–2425 (2014). https://doi.org/10.1021/sc500439w
10. Radošević, K., Cvjetko Bubalo, M., Gaurina Srček, V., Grgas, D., Landeka Dragičević, T., Redovniković, R.I.: Evaluation of toxicity and biodegradability of choline chloride based deep eutectic solvents. Ecotoxicol. Environ. Saf. **112**, 46–53 (2015). https://doi.org/10.1016/j.ecoenv.2014.09.034
11. Radošević, K., Ćurko, N., Gaurina Srček, V., Cvjetko Bubalo, M., Tomašević, M., Kovačević Ganić, K., Radojčić Redovniković, I.: Natural deep eutectic solvents as beneficial extractants for enhancement of plant extracts bioactivity. Lwt **73**, 45–51 (2016). https://doi.org/10.1016/j.lwt.2016.05.037
12. Mitar, A., Panić, M., Prlić Kardum, J., Halambek, J., Sander, A., Zagajski Kučan, K., Radojčić Redovniković, I., Radošević, K.: Physicochemical properties, cytotoxicity, and antioxidative activity of natural deep eutectic solvents containing organic acid. Chem. Biochem. Eng. Q. **33**(1), 1–18 (2019). https://doi.org/10.15255/CABEQ.2018.1454
13. Skulcova, A., Russ, A., Jablonsky, M., Sima, J.: The pH behavior of seventeen deep eutectic solvents. BioResources **13**(3), 5042–5051 (2019). https://doi.org/10.15376/biores.13.3.5042-5051
14. Hayyan, M., Looi, C.Y., Hayyan, A., Wong, W.F., Hashim, M.A.: In Vitro and in Vivo toxicity profiling of ammonium-based deep eutectic solvents. PLoS ONE **10**(2), 1–18 (2015). https://doi.org/10.1371/journal.pone.0117934

15. Dai, Y., van Spronsen, J., Witkamp, G.J., Verpoorte, R.: Natural deep eutectic solvents as new potential media for green technology. Anal. Chim. Acta **156**(4), 1701–1705 (2011). https://doi.org/10.1104/pp.111.178426
16. Liu, T., Liu, X., Spring, D. R., Qian, X., Cui, J., Xu, Z.: Quantitatively mapping cellular viscosity with detailed organelle information via a designed PET fluorescent probe. Sci. Rep. **4**(July) (2014). https://doi.org/10.1038/srep05418
17. Hayyan, M., Mbous, Y.P., Looi, C.Y., Wong, W.F., Hayyan, A., Salleh, Z., Mohd-Ali, O.: Natural deep eutectic solvents: cytotoxic profile. SpringerPlus **5**(1) (2016). https://doi.org/10.1186/s40064-016-2575-9
18. García, G., Aparicio, S., Ullah, R., Atilhan, M.: Deep eutectic solvents: Physicochemical properties and gas separation applications. Energy Fuels **29**(4), 2616–2644 (2015). https://doi.org/10.1021/ef5028873

MIX
Papier aus verantwortungsvollen Quellen
Paper from responsible sources
FSC® C105338
FSC
www.fsc.org

If you have any concerns about our products,
you can contact us on
ProductSafety@springernature.com

In case Publisher is established outside the EU,
the EU authorized representative is:
Springer Nature Customer Service Center GmbH
Europaplatz 3, 69115 Heidelberg, Germany

Printed by Libri Plureos GmbH
in Hamburg, Germany